Quantum Gravity in a Nutshell1

BALUNGI FRANCIS

Copyright ©Balungi Francis 2018,2019, 2020

First published 2018

Bill Stone Services

Balungi Francis asserts the moral right to be identified as the author of this work.

All rights reserved. Apart from any fair dealing for the purposes of research or private study or critism or review, no part of this publication may be reproduced, distributed, or transmitted in any form or by any means, including photocopying, recording, or other electronic or mechanical methods, or by any information storage and retrieval system without the prior written permission of the publisher.

Revised Edition: 2020

Cover by: znamide.

TABLE OF CONTENTS

Dedication   1

Preface   1

Solving Quantum Gravity   1

Singularity Resolution in Quantum Gravity   4

Planck stars   7

Singularity Resolution under the Assumption of Maximal Acceleration and Minimal length for both the Schwarzschild and Reissner-Nordstrom Black Hole   13

Maximal Acceleration in Quantum Gravity 15

Minimal Radius in Quantum Gravity   16

A Simple Derivation of the Minimum Radius of a Reissner - Nordstrom Black Hole: The Case of Accretion   19

Evidence for Minimal Length   23

Mass Ain't What It Used To Be   30

A Brief Account on the Implications of Quantum Gravity   37

Maximal acceleration 38

Minimal length   39

Minimal mass 39

Energy 40

Temperature 41

Hidden in plain sight1: A simple link between Quantum mechanics and General relativity 42

Einstein Field Equation for a Relationship between the DeBrogile Wavelength and the Energy Density of an Electromagnetic Wave 44

Derivation of the Schwarzschild-Hawking Power Law 46

Derivation of the Bekenstein –Hawking Area Entropy Law 48

Quantum Gravity in a Nutshell 51

The temperature of a black hole 53

Time taken by a black hole to evaporate 53

Entropy of a black hole 54

Thermal properties of solids 55

Stefan's Radiation Law 56

Hidden in plain sight2: From white dwarfs to Black Holes 58

Proof of the Chandrasker Mass Limit and the Lowest Principal Quantum Number from a New Approach to Quantum Gravity 59

The total Gravitational Binding Energy of a Star 60

The Theory of White -Dwarf Stars and Black Holes; The Limiting Mass at the Lowest Principal Quantum Number 63

What is Wrong With Hawking Temperature   66

Entropy of a Black Hole   68

On the Development of a Quantum Gravity-Hydrostatic Equation and its Implication to Physics-Minimum Black hole mass   69

Murder of Germans Sacred Cow: Experimental Test of General Relativity   75

What is Special about the Energy Density? 84

Application of the quantum energy density to space time singularities, information paradox, Planck stars, Emergence of the laws of Newton, galaxy rotation problem and the Tully-Fisher relationship   99

Derivation of Newton's law of gravity   102

Alternative Proof of Newton's Law of Universal Gravitation   105

The Tully-Fisher Relation   107

How to Calculate a Mysterious Repulsive Force Pulling Galaxies Apart   111

A Simple Link between MONDian Dynamics and the Dark Universe   118

A formula for apparent dark matter energy density in galaxies and clusters   121

A formula for the cosmological constant   123

Determination of the Hubble constant, Cosmological constant and the Vacuum Energy Density   126

Derivation of the Temperature and Entropy of Black Holes   134

Proof of the bekenstein-hawking black hole entropy law from first principles 137

Black hole Volume Entropy law 138

Particle Creation by Black Holes: Is it Hawking's Approach or My Approach? 141

Is There A Limit To How Small Black Holes Can Become? 157

On the Quantum Electrodynamics and Quantum Gravity Magnetic Field Limits. 162

Emergent Gravity 167

Gravity, Inertia and Electromagnetism as a result of Quantum vacuum fluctuations 167

Emergence of the laws of Newton 169

Force and Inertia 170

Newton's law of gravity 173

Emergence of electromagnetism 175

A New Alternative to Entropic Gravity 177

Derivation of force and inertia 179

Derivation of Newton's law of Gravity 180

Derivation of the law of electromagnetism 182

A new Approach to the Modification of Newtonian Dynamics (MOND) 184

Hypothesis 184

Calculation Of The Total Mass Of The Galaxy From Equation (d) 188

Reinventing Gravity "The Fifth Force"   191

The Modified Force Law   193

The Tully-Fisher Relation   194

Resolving the Proton Radius Puzzle   197

The Bekenstein Hawking Area-Entropy Law   203

First Method   204

Second Method   208

Temperature   209

Entropy   210

Third Approach   213

Temperature of a Black Hole   214

The Entropy of the Black Hole   216

The Time Taken by a Black Hole to Evaporate   217

Fourth Approach   220

Temperature of a black hole   220

Entropy of a black hole   222

New Physics: Regularization and Physics beyond the Standard Model   224

General Theory   228

A Grand Unification   238

The determination of the strength of the forces   240

Results   247

The unification of coupling calculations   247

The length scales at which the masses predicted by the standard model survive   250

The big bang acceleration and proton decay   251

A Unified Bohr and Quantum Gravity Theory   255

Everything   260

The formula for the quantization of quantum gravity   261

The energy equation 262

The mass equation   263

The maximal magnetic field 263

Time taken by a black hole to evaporate and its entropy   263

The quantum Hall Effect   264

Maximum Intensity   264

The Theory of Light 265

Photoelectric effect   269

Compton Effect   271

Making Sense with Semi-Classical Gravity 278

The meaning of semi-classical physics to an amateur   279

Applications of semi-classical physics   285

Radiation intensity of a black hole   285

The earliest period of time in the history of the universe   288

The Weidmann Franz- Lorenz law   291

The Art of Reductionism   293

onstruction of a Consistent Physical Theory of Nature   296

Is It Possible That There Is A Universe In Every Particle? 306

Newton's Biggest Blunder: Re-defining Gravity 309

Rule 1 309

Orbits around the Sun 309

Of The Galactic And Atomic 311

Rule 2 312

The modified Newton's universal law of gravitation (MONG) 313

Space-time Singularity or Quantum Black Holes? 315

What is real? Is it Volume or Area Entropy Law of Black Holes? 325

Is it Dark Matter, MOND or Quantum Black Holes? 331

What is real? General Relativity or Quantum Gravity 337

Derivation of the Energy density stored in the Electric field and Gravitational Field 350

Emergence of Gravity 353

Introduction 353

Determining the length scale at which the force of Gravity is strong between any two electrons 359

Hypothesis 359

Revised Gravitation Theory for the Modified Newtonian Dynamics (MOND) Paradigm and Beyond 365

Introduction 365

The Modified Force Law 365

The Tully-Fisher Relation   366

Graviton-Photon energy   367

Derivation of the Temperature of a Black Hole from the new force law   369

Epilogue   372

Glossary   375

Bibliography   400

Acknowledgments   419

About the Author   420

Dedication

To my wife Wanyana Ritah,

My sons Odhran & Leander

Preface

There is a need for a book on a Quantum Theory of Gravity that is not directed at specialists but, rather, sets out the concepts underlying this subject for a broader scientific audience and conveys joy in their beauty. The author has written with this goal in mind, and has succeeded admirably. This wonderful and exciting book is optimal for physics graduate students and researchers. The physical explanations are exceedingly well written and integrated with formulas. Quantum Gravity is the next big thing and this book will help the reader understand and use the theory.

Author's Note

Our search for ultimate understanding -- the Quantum Theory of Gravity -- has long been the quest of such great scientists as Aristotle, Newton, Einstein, Hawking and many others, and is expected to transform science, providing clarity and understanding that is unknown today, ideally via one single overlooked principle in nature. So far, this quest has produced theories such as Special Relativity, General Relativity and Quantum Mechanics, and such recent proposals as "Dark Matter" and "Dark Energy" in cosmology. Yet these all suffer serious internal problems and compatibility issues with each other, introducing even more questions, mysteries and paradoxes -- and often even violations of our laws of physics upon closer examination. As a result, the Quantum Theory of Gravity continues to elude us, leaving a fractured and divided scientific community with no clear direction forward. This has also resulted in the mathematisation of physics which has resulted in the reduction of the cosmos to a mathematical entity, which has not only confused physicists but accounts for their worst and most distracting assertions. This book makes a first case for the latter, with clear discussions exposing the flaws in the above concepts and more, while stepping back to take a good look at the scientific legacy we have inherited.

We are probably asking the wrong questions at the moment, nevertheless it is impossible to resist the temptation to try. After all, the other fundamental forces – except gravity – fit very neatly with quantum mechanics.

Balungi Francis 2018

Solving Quantum Gravity

The development of a quantum theory of gravity began in 1899 with Max Planck's formulation of "Planck scales" of mass, time, and length. During this period, the theories of quantum mechanics, quantum field theory and general relativity had not yet been developed. This means that Planck himself had no idea about what he had just developed-behind the Black board. Planck was not aware of quantum gravity and what it would mean for physicists. But he had just coined in formula one of the starting point for the holy grail of physics.

After P.Bridgman's disapproval of Planck's units in 1922, Albert Einstein having published the General Relativity theory, a few months after its publication he noted that "to the intra-atomic movement of electrons, atoms would have to radiate not only electromagnetic but also gravitational energy if only in tiny amounts, as this is hardly true in nature, it appears that quantum theory would have to modify not only Maxwellian electrodynamics, but also the new theory of gravitation". This showed Einstein's interest in the unification of Planck's quantum theory with his newly developed theory of Gravitation.

Then in 1933 came Bronstein's cGh-plan as we know it today. In his plan he argued a need for Quantum

Gravity. In his own words he stated: "After the relativistic quantum theory is created, the task will be to develop the next part of our scheme that is, to unify quantum theory (h), special relativity (c) and the theory of gravitation (G) into a single theory". Thus the theory of quantum gravity is expected to be able to provide a satisfactory description of the microstructure of space time at the so called Planck scales, at which all fundamental constants of the ingredient theories, c (speed of light), h (Planck constant) and G (Newton's constant), come together to form units of mass, length and time.

Therefore the need for the theory of quantum gravity is crucial in understanding nature, from the smallest to the biggest particle ever known in the universe. For example, "we can describe the behavior of flowing water with the long- known classical theory of hydrodynamics, but if we advance to smaller and smaller scales and eventually come across individual atoms, it no longer applies. Then we need quantum physics just as a liquid consists of atoms" Daniel Oriti in this case imagines space to be made up of tiny cells or atoms of space and a new theory of quantum gravity is required to describe them fully.

The aim of this book is to develop a theory capable of explaining the quantum behavior of the gravitational fields and thereafter explain the problems involving a combination of very high energy and very small dimensions of space such as, the behavior of Black holes and the study of the properties of the early universe.

For us to solve quantum gravity (QG), we need to address, understand and resolve in detail the problems brought about by the failure of the general theory of relativity (GR). Below I outline briefly where GR breaks down and later I resolve each of these problems with applications.

(1) General relativity fails to explain details near or beyond space-time singularities. That is, for high or infinite densities where matter is enclosed in a very small volume of space. Abhay Ashtekar says that; when you reach the singularity in general relativity, physics just stops, the equations break down

(2) General relativity fails to account for dark matter.

(3) General relativity also fails to be quantized.

Singularity Resolution in Quantum Gravity

The demand for consistency between a quantum description of matter and a geometric description of spacetime, as well as the appearance of singularities (where curvature length scales become microscopic), indicate the need for a full theory of quantum gravity. For example; for a full description of the interior of black holes, and of the very early universe, a theory is required in which gravity and the associated geometry of space-time are described in the language of quantum physics. Despite major efforts, no complete and consistent theory of quantum gravity is currently known, even though a number of promising candidates exist.

The first step towards the development of a quantum theory of gravity lies in studying the kind of physics behind black holes which are born when normal stars die or which were formed in regions of high energy density in the early universe. Black holes on the other hand, are completely collapsed stars that is, stars that could not find any means to hold back the inward pull of gravity and therefore collapse to a singularity.

This section is aimed at answering questions like; i) Do objects continually collapse to a singularity or there is a limiting distance below which the very notions of space and length cease to exist?

Theorem:- A star more than three times the size of our Sun collapses in this way; the gravitational forces of the entire mass of a star overcomes the electromagnetic forces of individual atoms and so collapse inwards. If a star is massive enough it will continue to collapse creating a Black hole, where the whopping of space time is so great that nothing can escape not even light, it gets smaller and smaller. The star in fact gets denser as atoms even subatomic particles literally get crashed into smaller and smaller space, and its ending point is of course a space time singularity.

The appearance of singularities in any physical theory is an indication that either something is wrong or we need to reformulate the theory itself. Singularities are like dividing something by zero. One such theory plagued by singularities is the General theory of relativity (GR) and the problems in GR arise from trying to deal with a universe that is zero in size (infinite densities). However, quantum mechanics suggests that there may be no such thing in nature as a point in space-time, implying that space-time is always smeared out, occupying some minimum region. The minimum smeared-out volume of space-time is a profound property in any quantized theory of gravity and such an outcome lies in a widespread expectation that singularities will be resolved in a quantum theory of gravity. This implies that the study of singularities acts as a testing ground for quantum gravity.

Loop quantum gravity (LQG) suggests that singularities may not exist. LQG states that due to quantum gravity effects, there must be a minimum distance beyond which the force of gravity no longer continues to increase as the distance between the masses become shorter or alternatively that interpenetrating particle waves mask gravitational effects that would be felt at a distance. It must also be true that under the assumption of a corrected dynamical equation of LQ cosmology and brane world model, for the gravitational collapse of a perfect fluid sphere in the commoving frame, the sphere does not collapse to a singularity but instead pulsates between a maximum and minimum size, avoiding the singularity.

Additionally, the information loss paradox is also a hot topic of theoretical modeling right now because it suggests that either our theory of quantum physics or our model of black holes is flawed or at least incomplete. and perhaps most importantly, it is also recognized with some prescience that resolving the information paradox will hold the key to a holistic description of quantum gravity, and therefore be a major advance towards a unified field theory of physics.

The paradox, as formulated, arises from considerations of the ultimate fate of the information that falls into a black hole: does it disappear as it falls into the black hole singularity? As well, what happens to the information of a black hole when it evaporates to nothing due to Hawking radiation? If a black hole loses all of its energy, then all of the information about all of the particles that fell in it would be lost as well. Of course the disappearance of information would be a violation of conservation laws of energy, which states that no energy or information can be destroyed.

Planck stars

To resolve the black hole singularities and the information paradox. We consider the possibility that the energy of a collapsing star and any additional energy falling into the hole could condense into a highly compressed core with density of the order of the Planck density. If this is the case, the gravitational collapse of a star does not lead to a singularity but to one additional phase in the life of a star: a quantum gravitational phase where the gravitational attraction is balanced by a quantum pressure.

Since the energy density or pressure is expressed as force per unit surface area of a star we have,

$$\rho = \frac{F}{A}$$

Therefore nature appears to enter the quantum gravity regime when the energy density of matter reaches the Planck scale. The point is that this may happen well before relevant lengths become planckian. For instance, a collapsing spatially compact universe bounces back into an expanding one. The bounce is due to a quantum-gravitational repulsion which originates from the modified Heisenberg uncertainty, and is akin to the force that keeps an electron from falling into the nucleus. The above given statement is based on the following facts:

The resolution of classical singularities under the assumption of a maximal acceleration has been studied using canonical methods for Rindler, Schwarzschild, Reissner-Nordstrom, Kerr-Newman and Friedman-Lemaitre metrics.

To reconcile quanum mechanics with general relativity, we develop a quantum geometry in relativistic phase space (Rindler space) in which the maximal (proper) acceleration of a particle is modified to read,

$$a = \frac{c^2}{2r}\alpha^n$$

Where, c is the constant speed of light, r is the linear dimension of a particle, α is the coupling constant (or size of the extra dimensions), n is a positive number (or the extra dimension number and **α^n** is the flux in the extra dimension

This acceleration is based on an assumption, that particles are extended objects, never to be identified with mathematical points in ordinary space. *This acceleration is important because it cures strong singularities that plague general relativity.* This acceleration is also a straight forward consequence of our modified uncertainty relation given as,

$$\Delta p \Delta r \geq \frac{\hbar}{2}\alpha^n, \Delta E \Delta t \geq \frac{\hbar}{2}\alpha^n$$

Where r represents the size of a star, in this case-horizon radius, p is the momentum of a particle approaching or falling into the hole of a star, α is the coupling constant and n is positive. From the above given uncertainty principle, we derive the planck length. such that when the momentum $\Delta p = mc$, the gravitational coupling constant for gravitational interactions is $\alpha = \frac{Gm^2}{\hbar c}$ and finally n=1/2. We get the planck length as the minimum length of space-time as

$$\Delta r = \left(\frac{\hbar G}{4c^3}\right)^{1/2}$$

Therefore from the uncertainity principle, the repulsion force is given by,

$$F = \frac{mc^2}{2r}\alpha^n$$

Therefore bounce does not happen when the universe is of planckian size, as was previously expected; it happens when the matter energy density reaches the Planck density in this way,

Let the surface area of a star be, $A = 4\pi r^2$ then the matter energy density will be given as,

$$\rho = \frac{mc^2}{8\pi r^3} \alpha^n$$

For a Schwarzschild black hole with radius $r = \frac{2GM}{c^2}$ and $\alpha = \frac{Gm^2}{\hbar c}$. We have a maximum energy density value wnen n=1 given as,

$$\rho = \frac{c^7}{\pi \hbar (8G)^2}$$

At this energy density, a Planck star is formed. The key feature of this theoretical object is that this repulsion arises from the energy density, not the Planck length, and starts taking effect far earlier than might be expected. This repulsive 'force' is strong enough to stop the collapse of the star well before a singularity is formed, and indeed, well before the Planck scale for distance. Since a Planck star is calculated to be considerably larger than the Planck scale for distance, this means there is adequate room for all the information captured inside of a black hole to be encoded in the star, thus avoiding information loss.

The analogy between quantum gravitational effects on Cosmological and black-hole singularities has been exploited to study if and how quantum gravity could also resolve the $r = 0$ singularity at the center of a collapsed star, and there are good indications that it does. For example, from the modified uncertainty principle, when the momentum of a particle or matter falling into a black hole is Planckian $p = m_{pl}c$ where m_{pl} is the Planck mass, we have,

$$r = \alpha^n l_p$$

Where l_p is the Planck length. Taking $\alpha = \frac{Gm^2}{\hbar c} = \left(\frac{m}{m_{pl}}\right)^2$ we have the size of a star as,

$$r = \left(\frac{m}{m_{pl}}\right)^{2n} l_p$$

Where m is the mass of the star and n is positive. For instance, if n = 1/6, a stellar-mass black hole would collapse to a Planck star with a size of the order of 10^{-10} centimeters. This is very small compared to the original star in fact, smaller than the atomic scale but it is still more than 30 orders of magnitude larger than the Planck length. This is the scale on which we are focusing here. The main hypothesis here is that a star so compressed would not satisfy the classical Einstein equations anymore, even if huge compared to the Planck scale. Because its energy density is already planckian.

Singularity Resolution under the Assumption of Maximal Acceleration and Minimal length for both the Schwarzschild and Reissner-Nordstrom Black Hole

Under the assumption of $\mu = m\alpha^{1/2}$ (where α is the coupling constant), in the Caianeillo maximum acceleration model ($A_{max} = \frac{\mu c^2}{m\lambda}$) , we derive the minimum radius to which a gravitating body can collapse in the commoving frame for both the Schwarzschild and Reissner-Nordstrom Black hole.

In the context of a geometrical unification of quantum mechanics and general relativity in phase space, Caianiello was the first person to propose the existence of a maximal proper acceleration for massive particles. Caianiello was able to derive the value $A_{max} = \frac{2mc^3}{\hbar}$ (1) for the maximum acceleration of a particle of rest mass m from the time-energy uncertainty relation. Caianiello model was based on two assumptions; $\hbar = \lambda\mu c$ and $\mu = 2m$ (2) for $A_{max} = \frac{\mu^2 c^3}{m\hbar} = \frac{c\hbar}{m\lambda^2} = \frac{\mu c^2}{m\lambda}$ (3).

Applications of Caianiello's model include cosmology, the dynamics of accelerated strings, neutrino oscillations and the determination of a lower neutrino mass bound. There is also evidence for maximal acceleration and singularity resolution in covariant loop quantum gravity found by Rovelli and Vidotto.

In this book we propose an adhoc assumption of $\mu = m\alpha^{1/2}$ where α is the coupling constant. This differs from Caianiello's model assumption of $\mu = 2m$. Therefore the maximum acceleration(3) will be given by,

$$A_{max} = \frac{c^2}{r} \alpha^{1/2} \qquad (4)$$

Where, r is the smallest possible distance between any two masses. In this book r takes values for the Schwarzschild and Reissner-Nordstrom radius.

Equation (4) given above reduces to the value $A_{max} = \frac{2mc^3}{\hbar}$ that was earlier derived by Caianiello under two conditions;

(i) When $r = \frac{2GM}{c^2}$ and $\alpha = 16\alpha_g^2$ for a Schwarzschild Black hole of mass M. Where α_g is the gravitational coupling constant $\frac{GM^2}{\hbar c}$.

(ii) When $r = \left(\frac{Ge^2}{4\pi\varepsilon_0 c^4}\right)^{1/2}$ and $\alpha = 4\alpha_g\alpha_e$ for a Reissner-Nordstrom Black hole. Where α_e is the electromagnetic coupling constant $\frac{e^2}{4\pi\varepsilon_0 \hbar c}$.

Maximal Acceleration in Quantum Gravity

Considering the event horizon of a Reissner-Nordstrom black hole of radius $r = \left(\frac{Ge^2}{4\pi\varepsilon_0 c^4}\right)^{1/2}$ and gravitational coupling $\alpha = \frac{GM^2}{\hbar c}$. Then substituting in (4), the growing acceleration approaching a classical singularity in the Reissner-Nordstrom metric is bounded by the existence of a maximal acceleration of;

$$a_{max} = \frac{M}{e}\left(\frac{4\pi\varepsilon_0 c^7}{\hbar}\right)^{1/2} \quad (5)$$

Where e is charge on an electron, ε_0 is the permittivity of free space and $\hbar$ is the reduced Planck constant.

Considering the event horizon of a Schwarzschild black hole of radius $r = \frac{GM}{c^2}$ and gravitational coupling $\alpha = \frac{GM^2}{\hbar c}$. Then substituting in (4), the growing acceleration approaching a classical singularity in the Schwarzschild metric is bounded by the existence of a maximal acceleration of;

$$a_{max} = \left(\frac{c^7}{G\hbar}\right)^{1/2} \quad (6)$$

Minimal Radius in Quantum Gravity

Because of the equivalence principle in the case of gravitational interaction, we propose to show here that the existence of a minimal length for both a Reissner and Schwarzschild Black hole is a straight forward consequence of our maximal acceleration value (4). In Newtonian law (center of mass system)

$$\frac{GM}{R^2} = \frac{c^2}{r} \alpha^{1/2}$$

Where, R is the radius of a Black hole (In this case the minimum radius to which a central mass will collapse), On arranging we have,

$$R = \frac{1}{\alpha^{1/4}} \left(\frac{R_s r}{2}\right)^{1/2} \qquad (7)$$

Where R_s is the Schwarzschild radius $R_s = \frac{2GM}{c^2}$.

Two results are thus deduced;

1) For $r = \left(\dfrac{Ge^2}{4\pi\varepsilon_0 c^4}\right)^{1/2}$ the radius of the event horizon of a Reissner Black hole and $\alpha = \dfrac{GM^2}{\hbar c}$, the minimum radius to which a gravitating body will collapse in a commoving frame of the Reissner-Nordstrom metric will have a value;

$$R_{min} = \left(\dfrac{\hbar e^2 G^2}{4\pi\varepsilon_0 c^7}\right)^{1/4} = \alpha_e^{1/4} l_p . \qquad (8)$$

Where l_p is the Planck length and α_e is the fine structure constant $1/137$.

$$R_{min} = 4.724 \times 10^{-36} m$$

2) Similarly, for $r = \dfrac{GM}{c^2} = \dfrac{R_s}{2}$ the radius of the event horizon of a Schwarzschild Black hole and $\alpha = \dfrac{GM^2}{\hbar c}$, the minimum radius to which a gravitating body will collapse in a commoving frame of the Schwarzschild metric will have a value;

$$R_{min} = (M)^{1/2} \left(\dfrac{\hbar G^3}{c^7}\right)^{1/4} . \qquad (9)$$

The above derivation clearly provides evidence for the existence of a maximal acceleration and minimal length which are both expected in the theory of quantum gravity to cure strong singularities such as, big bang, big crunch, black holes etc.

A Simple Derivation of the Minimum Radius of a Reissner-Nordstrom Black Hole: The Case of Accretion

Matter falling onto somebody is termed accretion. Suppose the matter is falling onto a star of mass M and radius R. Falling freely, it gains kinetic energy E_k in exchange for gravitational potential energy E_p. For a mass m falling from infinity to a distance r from the central mass M where relativistic quantum effects are taken into account, the E_k matches the E_p as

$$E_k = E_p$$

As the particle orbits closer and closer into a huge gravitational field its velocity increases up to a speed of light c, where the usual known kinetic energy formula does not apply. Instead we are forced to introduce a new formula that takes into account the gravitational coupling constant α_g as

$$E_k = \alpha_g^{1/2} mc^2 \qquad (10)$$

The self gravitation force of a star of radius R and mass M is known from Newton's gravitational force formula however the potential gravitational energy of a particle m falling from infinity to a distance r from a star will differ from the usual known potential formula as

$$E_p = \left(\frac{GMm}{R^2}\right) r \qquad (11)$$

Then surely,

$$\alpha_g^{1/2} mc^2 = \left(\frac{GMm}{R^2}\right) r$$

On cancelling like terms we have,

$$R = \frac{1}{\alpha_g^{1/4}} \left(\frac{R_s r}{2}\right)^{1/2} \qquad (12)$$

Where, $R_s = \frac{2GM}{c^2}$ is the Schwarzschild radius of a gravitating body and $\alpha_g = \frac{GM^2}{\hbar c}$ is the gravitational coupling constant that determines the strength of the gravitational force and G is the gravitational constant.

The mass eventually hits the surface of the star and its Kinetic energy is released as heat, and appears in some form of radiation. The radius of a star can then be determined using the above formula as: For a particle at the event horizon of a Reissner-Nordstrom Black hole, $r = \left(\frac{Ge^2}{4\pi\varepsilon_0 c^4}\right)^{1/2}$. Where e is charge on an electron, ε_0 is the permittivity of free space and $\hbar$ is the reduced Planck constant. The radius of this star is;

$$R = \left(\frac{\hbar e^2 G^2}{4\pi\varepsilon_0 c^7}\right)^{1/4} = 4.717444838 \times 10^{-36} \text{m}.$$

Then in terms of the Planck length we have,

$$R = 0.2923 l_p$$

Where l_p is the Planck length $l_p = \left(\frac{\hbar G}{c^3}\right)^{1/2} = 1.6144 \times 10^{-35}$ m.

Taking fourth powers on both sides of the equation we have,

$$R = \alpha_e^{1/4} l_p.$$

Where $\alpha_e = 1/137$ is the fine structure constant.

Therefore the above derivation implies that the radius of a Reissner-Nordstrom Black Hole is quantized in units of the Planck length and takes on only discrete units implying the quantized nature of space. In conclusion nature permits the existence of a minimum length beyond which the very notions of space and time cease to exist. I hope in my own view that this analysis will be useful for researchers involved in the field of quantum gravity and loop quantum cosmology.

Evidence for Minimal Length

General relativity predicts two kinds of singularities; the cosmological singularity at the beginning of our universe and the singularities at the centre of black holes. However, singularities signal the breakdown of general relativity and it is generally believed that they will be removed in a more fundamental theory of quantum gravity. The resolutions of singularities have been carried out directly in the frame work of Loop quantum gravity under the assumption of a maximal acceleration using canonical methods. However, in this example, singularities are resolved under the assumption of minimal length by creating a new cosmological model for the study of the gravitational collapse of a perfect fluid sphere. Two results are deduced;(i) a commoving observer accelerating with respect to his neighbors in a Reissner-Nordstrom space-time geometry will have a horizon at a distance bounded by a minimal value limit $R_{min} = \alpha_e^{1/4} l_p$ (Where l_p is the Planck length and α_e is the fine structure constant $1/137$) and (ii), a commoving observer accelerating with respect to his neighbors in a Schwarzschild space-time geometry will have a horizon at a distance bounded by a minimal value limit $R_{min} = (2M)^{1/2} \left(\frac{\hbar G^3}{c^7}\right)^{1/4}$. Therefore the generic bound on length and acceleration implies that the resolution of singularities is general and must be taken seriously.

Here we consider the gravitational collapse of a perfect fluid sphere- a gravitating body of mass M and radius R. Then for a test particle or an observer falling freely from infinity to a distance R_0 from the gravitating body, the spherically symmetric solution to the Einstein field equation will be given by; $R^2 = \frac{1}{\alpha_g^{1/2}} \left(\frac{R_s R_0}{2} \right)$ (13)

Where, $R_s = \frac{2GM}{c^2}$ is the Schwarzschild radius of a gravitating body and $\alpha_g = \frac{GM^2}{\hbar c}$ is the gravitational coupling constant that determines the strength of the gravitational force, G is the gravitational constant and c is the constant speed of light.

Therefore a commoving observer accelerating with respect to his neighbors in a given space- time geometry will have a horizon at a distance $R = \frac{1}{\alpha_g^{1/4}} \left(\frac{R_s R_0}{2} \right)^{1/2}$ bounded by a minimal value limit R_{min}. Correspondingly, the growing acceleration approaching a classical singularity is bounded by the existence of a maximal acceleration

$$a = \frac{c^2}{R_0} \alpha_g^{1/2}.$$

The existence of minimal length and maximum acceleration is of course something long expected in the quantum theory of gravity. Below we derive two important results for minimum radius and maximum acceleration supporting the results in loop quantum cosmology and black holes.

(i) Considering a test particle at the event horizon in the Reissner-Nordstrom metric (RN), $R_0 = R_{RN} = \left(\frac{Ge^2}{4\pi\varepsilon_0 c^4}\right)^{1/2}$. Where e is charge on an electron, ε_0 is the permittivity of free space and $\hbar$ is the reduced Planck constant. The minimum radius (size) to which a gravitating body can collapse in a commoving frame is;

$$R_{min} = \left(\frac{\hbar e^2 G^2}{4\pi\varepsilon_0 c^7}\right)^{1/4} = 4.717444838 \times 10^{-36} m.$$

This also implies a maximum acceleration of $a_{max} = \frac{M}{e}\left(\frac{4\pi\varepsilon_0 c^7}{\hbar}\right)^{1/2}$. Then in terms of the Planck length we have, $R_{min} = 0.2923 l_p$ (14) Where l_p is the Planck length $l_p = \left(\frac{\hbar G}{c^3}\right)^{1/2} = 1.6144 \times 10^{-35} m$. Taking fourth powers on both sides of the equation we have, $R_{min} = \alpha_e^{1/4} l_p$ (15). Where $\alpha_e = 1/137$ - the fine structure constant.

(ii) Considering a test particle at the event horizon in the Schwarzschild metric, $R_0 = R_S = \frac{2GM}{c^2}$. The minimum radius to which a gravitating body can collapse in a commoving frame is; $R_{min} = \left(\frac{4\hbar G^3 M^2}{c^7}\right)^{1/4}$. This also implies a maximal acceleration of,

$$a_{max} = \left(\frac{c^7}{4G\hbar}\right)^{1/2}.$$

Therefore a commoving observer accelerating with respect to his neighbors in a Reissner-Nordstrom space-time geometry will have a horizon at a distance bounded by a minimal value limit $R_{min} = \alpha_e^{1/4} l_p$. Where l_p is the Planck length and α_e is the fine structure constant 1/137. Correspondingly, the growing acceleration approaching a classical singularity in this metric is bounded by the existence of a maximal acceleration $a_{max} = \frac{M}{e}\left(\frac{4\pi\varepsilon_0 c^7}{\hbar}\right)^{1/2}$ where M is mass.

Also, a commoving observer accelerating with respect to his neighbors in a Schwarzschild space-time geometry will have a horizon at a distance bounded by a minimal value limit $R_{min} = (2M)^{1/2}\left(\frac{\hbar G^3}{c^7}\right)^{1/4}$. Correspondingly, the growing acceleration approaching a classical singularity in this metric is bounded by the existence of a maximal acceleration $a_{max} = \left(\frac{c^7}{4G\hbar}\right)^{1/2}$.

It has been deduced in (i) above that, the resolution of singularities occurs as a result of a fundamental discreteness of space. This is based on the fact that the minimum radius or size is proportional to the Planck length l_p (14). This is one of the promising results of this essay. The presence of l_p implies a discreteness of space or length spectra which is manifested by the presence of the fine structure constant (15). However, in (ii) singularities are avoided in a limit $M = M_p$ (Planck mass), by imposing a minimum length $R_{min} = 2^{1/2} l_p$. Therefore the generic bound on length and acceleration implies that the resolution of singularities is general and must be taken seriously.

Unlike other models, the cosmological model (13) created in this example directly predicts a limit on the length and acceleration, thus providing evidence for the resolution of the classical singularity. The derivation in (i) is unique in that the value of the fine structure constant comes out as a direct result of the theory, which has never been witnessed in any promising theory of quantum gravity, not even in LQG or string theory.

Remark: In a more general form we can express (13) as, $R^2 = \frac{1}{\alpha_g{}^s}\left(\frac{R_s R_o}{2}\right)$, where s= 0,1,2,3,......,1/2, Such that when s=0, $R = \left(\frac{R_s R_o}{2}\right)^{1/2}$. What name should be given to s is left for the reader to decide. However we can denote s as an extra dimension number.

We have clearly modified the geometry of Rindler space by the introduction of the coupling $\alpha^{1/2}$ into the formula for acceleration. We have witnessed that the presence of $\alpha^{1/2}$ into the formula for acceleration leads to an exact evidence for the existence of the maximal acceleration and minimal length for both the Reissner-Nordstrom and Schwarzschild black holes in quantum gravity. The split horizon in a Rindler wedge at a distance $R= c^2/a$ for the acceleration a has been modified here, hope you have witnessed how $\alpha^{1/2}$ changes all of this. This implies that there is some fundamental limitation on how much acceleration a particle could experience based on the strong-field behavior of the fundamental force causing it.

Results of the maximal acceleration and minimal length for the Reissner Black hole have not been derived anywhere in literature. These clearly impose a general bound on acceleration and length (in Reissner space time geometry) with implications. For example, a black hole the size of an electron ($m_e = 9.11 \times 10^{-31}$kg), imposes an acceleration of $a_{max} = 2.732 \times 10^{30}$m/s^2. So this accelerated frame would detect a Unruh radiation at 1.1×10^{10}K. Also the minimal length result implies the existence of the discreteness (granular nature) of space and cures the singularities that plague General relativity by imposing a general bound on length of 4.724×10^{-36}m.

In conclusion, a corrected Rindler space geometry directly proves an existence for the maximal acceleration and minimal length in quantum gravity, not only for the Schwarzschild metric with a horizon distance half of the Schwarzschild radius but also for the Reissner metric. Therefore the introduction of $\alpha^{1/2}$ in the formula for acceleration must be thoughtfully investigated as this solves all the problems brought about by the General relativity theory.

Mass Ain't What It Used To Be

The origin of mass problem is at the forefront of those big unsolved problems in the standard model of physics. My first insight about mass came in 2000 in a lecture about Newton's mechanics probably about the study of Newton's second law of motion. The problem is important to me because the primary role of mass is to mediate gravitational interaction between bodies, and no theory of gravitational interaction reconciles with the currently popular standard model of particle physics. But because the problem started with inertia, we again revisit Newton's law to create a model through which all the masses of elementary particles can be generated.

Recall from the previous section, the modified Rindler space with an acceleration given by,

$$a = \frac{c^2}{r} \alpha^{1/2}$$

Because we desire to have a mass formula just in terms of universal interaction couplings (like e) and basic constants like G and $\hbar$. We are therefore propted to use the radius of a RN electrically charged black hole as this is purely made of only universal constants.

$$r = \left(\frac{Ge^2}{4\pi\varepsilon_0 c^4}\right)^{1/2}$$

Then inserting this radius into the formula for acceleration we obtain,

$$a = \frac{c^4}{e}\left(\frac{4\pi\varepsilon_0 \alpha}{G}\right)^{1/2}$$

Where, c is the constant speed of light, e is the charge on an electron, α is the dimensionless coupling constant, ε is the permittivity of free space and G is the universal gravitational constant.

We have therefore obtained the acceleration formula made of purely universal fundamental physical constants. Then from Newton's law of motion where force is the product of mass and acceleration, F=ma, we have

$$F_N = \frac{mc^4}{e}\left(\frac{4\pi\varepsilon_0 \alpha}{G}\right)^{1/2}$$

Having found the force on a particle due to its inertia, we would like also to deduce the force of attraction on the same particle. Assuming the self attraction of a particle of radius R to be caused by the quantum fluctuation of the vacuum, then the Casimir force will be given by,

$$F = \frac{\hbar c}{R^2}$$

Taking the radius of the particle to be of Schwarzischild radius $R = \frac{Gm}{c^2}$, then the force due to self attraction will be given by,

$$F_c = \frac{\hbar c^5}{G^2 m^2}$$

By the principle of equivalence $F_N = F_c$, this then gives the mass formula as,

$$m = \left(\frac{\hbar^2 c^2 e^2}{4\pi\varepsilon_0 G^3 \alpha}\right)^{1/6}$$

We have therefore created a mass formula that is made of purely fundamental physical constants. The next step is to test the theory to see if it actually makes fundamental sound predictions. First we would like to know if the above given mass formula gives the Planck mass value and what coupling constant makes this possible?

We find that, when the dimensionless physical constant is the fine structure constant or the electromagnetic coupling constant $\alpha_e = \frac{e^2}{4\pi\varepsilon_0 \hbar c}$, then the mass is probably the planck mass,

$$M_{pl} = \left(\frac{\hbar c}{G}\right)^{1/2}$$

Because the above result is general and true on cosmological grounds, then the other masses of elementary particles will be generated in a similar manner but on grounds that the Planck mass is the upper bound on mass. For example, for the case of the gravitational coupling constant, $\alpha = \frac{Gm^2}{\hbar c}$

$$m = \left(\frac{(\hbar c)^3 e^2}{4\pi\varepsilon_0 G^4}\right)^{1/8} = 0.54 M_{pl}$$

However things become complicated with the Higgs mass. The coupling constant required in calculating the Higgs mass of $2.2375 \times 10^{-25} kg$ according to the theory given above is enormous with a value given as $\alpha = 6.2 \times 10^{99}$. To show that the result given here is correct, we embark on the derivation of the life time of a star as,

From the modified uncertainity principle given in chapter1, the life time of the Higgs mass is here given by,

$$\Delta t = \frac{\hbar}{2\Delta E} \alpha^{1/2}$$

Where ΔE is the change in energy of the particle and according to Einstein mass-energy relation this energy is given as,

$$\Delta E = mc^2 = m = \left(\frac{\hbar^2 c^{14} e^2}{4\pi\varepsilon_o G^3 \alpha}\right)^{1/6}$$

For the Higgs boson particle, where the coupling $\alpha = 6.2 \times 10^{99}$, the binding energy will be calculated to be,

$$\Delta E = 2.0136 \times 10^{-8} J$$

Then on substitution into the life time formula above, we get the life time of the Higgs boson as,

$$\Delta t = 2.063 \times 10^{23} s$$

The above calculated lifetime agrees with experimental observations. This therefore proves that the value of the coupling constant for the higgs mechanism given here is correct.

Last but not least, when the dimensionless coupling constant is unity, that is $\alpha=1$. We obtain the following mass value,

$$m = 0.44 M_{pl}$$

In general, the coupling constant and generation of mass must follow this simple rule

$$\alpha = \left(\frac{M_{pl}}{m}\right)^6 \alpha_e$$

Such that when m=1.4Mpl, we get the value of the strong interaction gluon coupling (asymptotic freedom) as,

$$\frac{\alpha}{\alpha_e} = 0.13$$

We have shown above that the generation of mass is possible however it requires the determination of the exact number of the dimensionless coupling constant of the underlying theory. We have found that the coupling constant in the generation of the higgs mass is enormous and requires a small energy to probe it. We have also shown that the particles collapse in the same way as the stars do which calls for another branch of physics to probe this study efficiently. Therefore the study of the origin of mass is important in the study of the origin of the universe and the unification of all physics using the coupling constants for different fundamental laws.

A Brief Account on the Implications of Quantum Gravity

The existence of a maximal acceleration in quantum gravity has great implications for physicists venturing into the field of quantum gravity. Although it has been studied in the previous chapters we again bring it here using different methods for purposes of imposing general bounds on the acceleration, length, mass, energy and temperature. The study of maximal acceleration is wide and that is why we are going to spend a great deal of time here in analyzing the consequences of its existence. The problem is important in calculating various limiting cases for both the theory of quantum gravity (at the Planck epoch) and the theory of quantum electrodynamics. As we shall see, all the calculations undertaken in the process lead us to one thing, a consistent quantum theory of gravity.

To differ from Newton's laws of motion and Einstein's theory of general relativity, our acceleration will depend on the dimensionless coupling constant which determines the strength of the force in any given interaction as was described earlier

$$a_{acel} = \frac{c^4}{e}\left(\frac{4\pi\varepsilon_o \alpha}{G}\right)^{1/2}$$

Where, c is the constant speed of light, e is the charge on an electron, α is the dimensionless coupling constant, ε is the permittivity of free space and G is the universal gravitational constant.

Various examples have been given below in which the theories of quantum gravity and quantum electrodynamics will act as limiting cases,

Maximal acceleration

For example, where the quanta exchanged between two electrons is a photon in the case of the electromagnetic force we have the electromagnetic coupling constant or the fine structure constant as, $\alpha = \frac{e^2}{4\pi\varepsilon_0 \hbar c}$ which deduces the acceleration to, $a = \frac{c^{7/2}}{(\hbar G)^{1/2}}$. This is the allowed maximum acceleration for quantum gravitational effects at the Planck epoch. But for the case where the quanta exchanged between two electrons is a graviton for a gravitational force, we have the gravitational coupling constant as, $\alpha = \frac{Gm^2}{\hbar c}$ which gives the acceleration on a quantum electrodynamics scale as,

$$a = \frac{m}{e}\left(\frac{4\pi\varepsilon_0 c^7}{\hbar}\right)^{1/2}.$$

Minimal length

Then the minimal radius to which a gravitating body or an electron can collapse in a commoving frame can also be deduced as, If we equate Newton's law of universal gravitation to our newly developed force as, $\frac{Gm^2}{R^2} = \frac{mc^4}{e}\left(\frac{4\pi\varepsilon_0 \alpha}{G}\right)^{1/2}$ We obtain the area as, $R^2 = \frac{me}{c^4}\left(\frac{G^3}{4\pi\varepsilon_0 \alpha}\right)^{1/2}$ then for $\alpha = \frac{Gm^2}{\hbar c}$, as in the first example, we obtain the minimum radius for a charged particle for quantum gravitational effects as,

$R_{min\ 1} = (Ge)^{1/2}\left(\frac{\hbar}{4\pi\varepsilon_0 c^7}\right)^{1/4}$ =4.717444838 × 10^{-36}m, or $R_{min\ 1} = 0.2923 l_p$, where l_p is the Planck length. Then the fine structure constant will be calculated as, $\alpha = \left(\frac{R_{min\ 1}}{L_p}\right)^4 = \frac{1}{(3.42155)^4} = \frac{1}{137.054}$ Hence solving one of the unsolved problems in physics. But for $\alpha = \frac{e^2}{4\pi\varepsilon_0 \hbar c}$, we obtain the minimal radius due to torsion in the gravitational field as, $R_{min\ 2} = (m)^{1/2}\left(\frac{\hbar G^3}{c^7}\right)^{1/4}$

Minimal mass

On another note, we could derive the mass formula only if we equate the force $F = \frac{\hbar c^5}{G^2 m^2}$ to our force formula $F = \frac{mc^4}{e}\left(\frac{4\pi\varepsilon_0 \alpha}{G}\right)^{1/2}$, then the mass expression is deduced as $m = \frac{\hbar^{1/3} e^{1/3} c^{1/3}}{(4\pi\varepsilon_0)^{1/6} G^{1/2} \alpha^{1/6}}$ This gives the Planck mass at

$\alpha = \frac{e^2}{4\pi\varepsilon_0 \hbar c}$, Also the mass that incorporates all the constants of nature when $\alpha = \frac{Gm^2}{\hbar c}$ is deduced as,
$m = \left(\frac{(\hbar c)^3 e^2}{4\pi\varepsilon_0 G^4}\right)^{1/8} = 1.177535 \times 10^{-8}$ kg. This could be the mass of the graviton.

Energy

Then from Einstein's proposal for the radiation of the gravitational energy, we have an expression for energy as,
$W = ma_{acel}R = \frac{mc^4}{e}\left(\frac{4\pi\varepsilon_0 \alpha}{G}\right)^{1/2} R$, Where R is the radius of orbit of an electron around the nucleus of an atom, for $R \sim \frac{\hbar}{mc}$, and $\alpha = \frac{e^2}{4\pi\varepsilon_0 \hbar c}$ we obtain the maximum energy as, $W = \left(\frac{\hbar c^5}{G}\right)^{1/2}$. This is the Planck energy at the Planck epoch.

But for $R \sim \frac{\hbar}{mc}$ and $\alpha = \frac{Gm^2}{\hbar c}$ we obtain the energy possessed by an electron of mass m in the electromagnetic field as, $W = \frac{2m}{e}(\pi\varepsilon_0 c^5 \hbar)^{1/2}$, Then at the Planck epoch when $m = \sqrt{\frac{\hbar c}{G}}$ (the Planck mass), the energy required to accelerate an electron in the gravitational field will be given by, $W = \frac{2\hbar c^3}{e}\left(\frac{\pi\varepsilon_0}{G}\right)^{1/2}$. In the case where the energy of the quantized states of the hydrogen atom $\frac{me^4}{16\pi^2 n\hbar^2 \varepsilon_0^2}$, is equated to the energy of $\frac{2m}{e}(\pi\varepsilon_0 c^5 \hbar)^{1/2}$, we obtain a crucial relationship between the fine structure constant and the principal quantum number n, as $\alpha = \left(\frac{n}{\pi}\right)^{2/5}$. Then the smallest quantum number that will give the value of the fine structure constant will be given by, $n = 1.429101876 \times 10^{-5}$. This is the lower limit for the quantum theory.

Then on the quantum gravitational scale, in which the Bohr's quantized energy is equated to our energy $\frac{2\hbar c^3}{e}\left(\frac{\pi\varepsilon_0}{G}\right)^{1/2}$, we obtain the relationship between the principal quantum number n, the gravitational coupling constant α_G and the fine structure or the electromagnetic constant α_E as, $n^2 = \alpha_G \alpha_E^5$ or $n = \alpha_G^{1/2} \alpha_E^{5/2} \sim 3.493 \times 10^{-25}$. This value implies an upper bound on the energy states for a combined theory of gravity and quantum mechanics between two protons.

Temperature

Last but not least, we could write a modified Unruh-Davis effect as, $T = \frac{\hbar c^3}{k_B}\left(\frac{\varepsilon_0 \alpha}{\pi G}\right)^{1/2}$, when it is equated to the Hawking temperature effect $T = \frac{\hbar c}{k_B}(\Lambda)^{1/2}$, where Λ is the curvature of space, we obtain the curvature as, $\Lambda = \frac{c^4 \varepsilon_0}{\pi e^2 G}\alpha$, then for $\alpha = \frac{e^2}{4\pi\varepsilon_0 \hbar c}$, we obtain the maximum curvature for quantum gravitational effects as, $\Lambda = \frac{c^3}{4\pi^2 G\hbar}$. But when $\alpha = \frac{Gm^2}{\hbar c}$, we obtain the curvature for quantum electrodynamics effects as, $\Lambda = \frac{m^2 c^3 \varepsilon_0}{\pi e^2 \hbar}$.

Hidden in plain sight1: A simple link between Quantum mechanics and General relativity

The Einstein field equation is written in the form, $G_{\mu\nu} + \Lambda g_{\mu\nu} = \frac{8\pi G}{c^4} T_{\mu\nu}$ where, the expression on the left represents the curvature of space time while the expression on the right represents the matter-energy content of the universe.

Then assuming a quantum state in which a gravitating particle of radius R is acted upon by all classical forces, the expression on the left, the metric of space time curvature can be written in a special form as,

$$G_{\mu\nu} + \Lambda g_{\mu\nu} = \frac{\hbar c}{F_u R^4} \dots\dots\dots\dots\dots 16$$

Where, F_u is the force arising from a quantized field in zero point vacuum energy, $\hbar$ is the reduced planck constant and c is the constant speed of light.

Whereas the expression on the right (the stress-energy tensor) will be written in a form,

$$T_{\mu\nu} = \frac{F_G F_E}{F_u R^2} \dots\dots\dots\dots\dots\dots17$$

Where, F_G is the Newtonian classical gravitational force between two particles of mass m, F_E is the electrostatic force between two particles of charge q and G is the gravitational constant.

Then in terms of the pressure and energy density the stress-energy tensor is,

$$T_{\mu\nu} = P_g \varphi = \rho_E \varphi$$

Where $\varphi = \frac{F_E}{F_u}$ is the coupling of forces, P_g is the pressure ($P_g = \frac{GM^2}{R^4}$) and ρ_E is the energy density or the potential gravitational energy per unit volume R^3.

Then in its simple form, the Einstein Field equation may be expressed as,

$$\frac{\hbar c}{F_E R^4} = \frac{8\pi G}{c^4} P_g = \frac{8\pi G}{c^4} \rho_E \dots\dots\dots\dots18$$

From the above equation, the gravitational potential field is analogous to the quantum gravitational potential by

$$\nabla^2 \phi = 4\pi G \rho_E \text{ (Classical)}$$

$$\text{And, } \nabla^2 \phi = \frac{\hbar c^5}{2 F_E R^4} \text{ (quantum)}$$

We have coupled a quantum system to a classical one by simply denoting the metric of space time in Einstein's field equation as $\frac{\hbar c}{F_u R^4}$, this out come gives us a unique technique through which we can express the gravitational effects in terms of quantum mechanics.

Einstein Field Equation for a Relationship between the DeBrogile Wavelength and the Energy Density of an Electromagnetic Wave

In a cyclotron the acceleration of a particle describing circular motion at a distance R in a magnetic field B will be given as

$$a_x = \frac{2\pi B v R}{q \mu_0},$$

Where μ_0 is the permeability of free space

q is the charge on a particle and

v is a velocity at right angles to the direction of the field B

But when quantum and gravitational effects are taken into account, we are led to a different formula for the acceleration given by

$$a_y = \frac{\hbar c^5}{8\pi Gm\, R^2 Eq},$$

This is deduced from equation18 where, the electrostatic force on a charge q in vicinity of the electric field E is $F_E = Eq$, E=Bc and the inertial force is $F_G = ma_y$.

Then at a point where the two accelerations are equal that is, $a_x = a_y$, we are led to,

$$\frac{\lambda}{2\pi R^3} = \left(\frac{8\pi G}{c^4}\right)\frac{EB}{\mu_0 c} = \left(\frac{8\pi G}{c^4}\right)\rho$$

Where $\rho = \frac{EB}{\mu_0 c}$ is the energy density of an electromagnetic wave in vacuum and $\lambda = \frac{\hbar}{mv}$ is the deBrogile wave length

The formula obtained above is the solution to the Einstein field equation in which the wave properties of matter in terms of the DE Brogile wavelength are related to the wave properties of an electromagnetic wave in terms of the energy density of an electromagnetic wave. The expression on the left represents the quantum nature of wave mechanics while that on the right represents the classical nature of electromagnetic waves interrelated by the gravitational constant.

It is therefore true from our derivations that, when the classical acceleration of particles in the cyclotron is equal in magnitude to the modern acceleration (not yet observed), we deduce properties of a wave on both a quantum and classical realm simultaneously. This means that both the wave and particle properties of matter cannot be separated in any experiment and or observation, hence a wave –particle duality of matter

Derivation of the Schwarzschild-Hawking Power Law

Suppose a force F_E does work on a Black hole of mass M to move it through a small displacement Δd in time Δt, where $\Delta d / \Delta t$ is the average speed v, then the power is,

$$P = F_E v$$

But from equation 18, $F_E = \dfrac{\hbar c^5}{8\pi G^2 M^2}$, is the force required to displace or accelerate a black hole, the force increases as the mass of the black hole decreases.

If we let $v = c$ then the power of a Black hole will be given by,

$$P = \dfrac{\hbar c^6}{8\pi G^2 M^2}$$

This sets a limit to which velocity a black hole can be accelerated. Note: the above formula has not yet been derived in the frame work of semi classical gravity. If this is semi classical gravity, then we are towards achieving a quantum theory of gravity. I therefore leave the derivation above to the entire scientific community to investigate.

However the expression above differs from that deduced from the Stefan-Boltzmann radiation power law of

$$P = \dfrac{\hbar c^6}{15360\pi G^2 M^2}.$$

Meaning that, it requires a velocity of $1.5625 \times 10^5 m/s$ to obtain this power from our derivations.

Derivation of the Bekenstein –Hawking Area Entropy Law

The energy or work done by a black hole will to a great degree depend on the surface area of the event horizon A and on the Compton wavelength λ of a black hole provided the force exerted on this black hole remains a constant as,

$$W = F_E \frac{A}{\lambda}$$

The Compton wavelength is $= \frac{2\pi \hbar}{mc}$, and F_E is known, thus the energy is,

$$W = \frac{Ac^6}{16\pi^2 G^2 m}$$

This implies that, the energy of a black hole is proportional to the surface area of the event horizon but inversely proportional to its mass.

If we apply the above statement to entropy which is energy per unit temperature, S=W/T we can deduce the entropy of a black hole. Let us deduce the expression for temperature: when the electric force applied on a body of mass m through a schwarzichild's radius $R_s = \frac{2Gm}{c^2}$, results into an energy equal to the translational kinetic energy as, $F_E R_s = kT$, where k is the Boltzmann's constant. Then the expression for temperature will be given as, $T = \frac{\hbar c^3}{4\pi Gmk}$, this is the temperature of a black hole. Then substituting for the energy and temperature in the entropy formula we obtain,

$$S = \frac{Ac^3 k}{4\pi G \hbar} \dotfill 19$$

This is the entropy of a black hole in its simplest form.

In conclusion, the Book has presented a new approach to Quantum Gravity that is different from string theory and loop quantum gravity by Carlo Rovelli and Edward Witten. The major result of the research is the derivation of the Bekenstein-Hawking area entropy law from first principles using new methods with a well defined calculation where no infinities appear. As far as this book is concerned there is no other theory from which such a calculation can proceed. Hence the methods used in here are the only one from which a detailed quantum theory of gravity "Holy Grail of modern physics" precedes and where the result of the Bekenstein-Hawking area entropy law can be achieved.

Quantum Gravity in a Nutshell

It seems as though we must use sometimes the one theory and sometimes the other, while at times we may use either. We are faced with a new kind of difficulty. We have two contradictory pictures of reality; separately neither of them fully explains the phenomena of light, but together they do. Albert Einstein

In this section a new approach towards Quantum Gravity is presented, we try to study the theory from new assumptions which are far more different from models that have been used by scientists for centuries. Most physicists have clung to old models or complex mathematical scientific methods to explain phenomenon. They are trying to explain physics using the mathematics that was earlier used by Einstein, Richard P. Feyman etc. The mathematical ideas that were presented by these physicists were complicated and such has been difficult to understand and of course misleading (Read Lost in Math, a book by Sabine Hossenfielder). For example, in a statement by Dr Lee Smolin, "the mathematisation of physics has resulted in the reduction of the cosmos to a mathematical entity, which has not only confused physicists but accounts for their worst and most distracting assertions".

There is a wide spread speculation that the mathematical formulation of physics has not only confused physicists but has also lead to failures in the development of a quantum theory of gravity.

Physics as a subject should be simple and elegant, trying to explain everything from one source. In other words trying to explain all of physics from one equation call it "the principle of least action". Imagine deducing the equations of gravity, quantum mechanics, electromagnetism, heat etc from one equation, wouldn't it be unique than holding about ten books about a different subject of physics each starting from its own source?

The principle of least action in simple terms means; to understand how to get from point A to point B using the least amount of physical work for example taking an elevator rather than using the stairs, in otherwords deducing the most fundamental physical equations from one principle as we are yet to find out.

Assuming that the ratio of the gravitational force to the electric force is equal to the gravitational coupling constant, we then have,

$$[8\pi G/c^4][Ee][GM^2/c]=n^2\hbar \qquad 20$$

Where G– gravitational constant, n- quantum number, c- constant speed of light, e- charge on an electron, M –

mass and is the reduced Planck constant. Ee is the electromagnetic force, and $8\pi G/c^4$ is the gravitational force at the swcharzichild's radius. From the above given principle, we deduce the temperature of a black hole, the time taken by a black hole to evaporate, entropy of a black hole, the wiedemann franz law and the stefan's radiation law.

The temperature of a black hole

On arranging equation 20 to get the random transilational kinetic energy, we obtain

$$[GM/c^2]Ee = n^2 c^3 \hbar /8\pi GM = kT$$

Where k is the boltzmann's constant and T is the temperature. Hence at n=1,

$$T = c^3 \hbar /8\pi GMk \qquad 21$$

This is the known temperature of a black hole that was originally derived by Hawking

Time taken by a black hole to evaporate

On dividing eqn20 by the momentum Mc we obtain, the time t given by

$$t = Mc/Ee = 8\pi G^2 M^3 / n^2 \hbar c^4$$

Such that when n=0.03953

$$t = 5120\pi G^2 m_o^3 / \hbar c^4 \qquad 22$$

This is the known time of a black hole that was originally derived by Hawking

Entropy of a black hole

Squaring both sides of equation 20 and arranging we generate the intensity as

$$W/tA = E^2 e^2 / 2nh = n^3 c^{10} \hbar / 256\pi^3 G^4 M^4 \qquad 23$$

Where A is the area on which the radiations fall, W is energy, and t is time. But entropy is energy divided by temperature Eq so then

$$W/T = S = (n^3 c^{10} \hbar /256\pi^3 G^4 M^4)(tA/T)$$

Since t is known from Eq22 and T from Eq21, then at $n = \pi$

$$S = Akc^3 /4G\hbar$$

This is the known Bekenstein-Hawking area entropy law

Thermal properties of solids

From the intensity equation23,

$$E^2 e^2 /2nh = c^{10} \hbar /256\pi^3 G^4 M^4$$

Arranging the above equation to introduce in the translational kinetic energy obtained above [$kT = c^3 \hbar /8\pi GM$], we have

$$\pi M^2 G^2 E^2 /3c^4 = (\pi^2/3e^2)(c^3 \hbar /8\pi GM)^2 = (\pi^2/3e^2)T^2 k^2$$

Dividing both sides by T we obtain on the left hand side of the equation the ratio of thermal conductivity K to electric conductivity δ as

$$K/\delta = (\pi^2/3)(k/e)^2 T$$

This is the known wiedemann fanz law

Stefan's Radiation Law

Still from the intensity Eqn23 we can arrange the expression on the left hand side of the equation, to read as

$$W/tA = E^2e^2/2nh = (1.875n^3/\pi^2)(\pi^2/60h^3c^2)(c^3\hbar/8\pi GM)^4$$

This is the same as Eqn23 only that it is arranged to predict something. But at n=1 and $(c^3\hbar/8\pi GM) = kT$, the rate at which energy is radiated is given by

$$W/t = A(1.875/\pi^2)(\pi^2/60h^3c^2)(Tk)^4 = 0.19\sigma AT^4$$

Where $\sigma = \pi^2/60h^3c^2$ is the Stefan boltzmann's constant

In conclusion, I encourage further research into this field. In other words this could be a stepping stone towards the development of a theory of everything via a simpler path of least action.

Hidden in plain sight2: From white dwarfs to Black Holes

A precise and consistent quantum theory of gravity has not yet been proved, not even by the self proclaimed geniuses of this time. We are aware and satisfied that classical General Relativity is the most precise description of gravity due to its predictable nature. The left hand side of Einstein field equation represents the metric of space time curvature while the right hand side represents the matter - energy content of the classical matter fields of pressure and energy density. It is known that quantum mechanics plays an important role in the behaviour of the matter fields but has no place in the Einsteins field equations.According to S.W.Hawking (1975), one therefore has a problem of defining a consistent scheme in which the space time metric is treated classically but is coupled to the matter fields which are treated quantum mechanically.

In this book we propose that, in order to estimate stellar parameters to a high degree of accuracy for both microscopic and macroscopic descriptions of white dwarfs and black holes one has to treat the right hand side of Einstein field equation quantum mechanically as,

$$\left(\frac{8\pi G}{c^3}\right)^{3/2} \frac{m_H}{\hbar^{1/2}} P_{eg}$$

$P_{eg} = \frac{f_e f_g}{\hbar c}$, where P_{eg} is the total pressure, f_g is the gravitational force, f_e is the electric force , G is the gravitational constant, c is the constant speed of light, $\hbar$ is the reduced planck constant and m_H is the mass of an Hydrogen atom.

Proof of the Chandrasker Mass Limit and the Lowest Principal Quantum Number from a New Approach to Quantum Gravity

Although in the Bohr theory of an hydogen atom orbit quantization doesnot permit a lower orbit than the bohr radius of $a_o = 0.53\text{Å}$, this section sets out to show that this is not the case with white dwarfs due to the state of a hydrogen atom under high pressure.

We know from the Chandrasker derivations that, the equation governing the hydrostatic equilbrium of a star is given by

$$-r^2 P(r) = GM(r)\rho(r) \text{ or } r\frac{dP}{dr} = -\frac{GM(r)}{r^2}\rho$$

Where P denotes the total pressure, ρ is density, and M(r) is the mass interior to a sphere of radius r.

We could however write the same equation in a different form given by

$$P_{eg} r^2 = \frac{\hbar^{1/2} c^{9/2}}{(8\pi G)^{3/2} m_H}, \text{ where } P_{eg} = \frac{f_e f_g}{\hbar c} \qquad (24)$$

The total Gravitational Binding Energy of a Star

The electric potential energy $E_e = f_e r$ as we know it can be can be deduced from (24) and is given by,

$$E_e = \left(\frac{\hbar}{8\pi G}\right)^{3/2} \frac{c^{11/2}}{E_g m_H}$$

where E_g is the gravitational potential binding energy given by $f_g r$

Using the principle of energy equipartition, we assume that the electric binding energy is of order the discrete energy of an hydrogen atom from Bohrs theory as,

$$E_n = E_e \Rightarrow \left(\frac{\hbar}{8\pi G}\right)^{3/2} \frac{c^{11/2}}{E_g m_H} = \frac{m_H k_e^2 e^4}{2n^2 \hbar^2}$$

where, k_e is the Coulomb constant, e is the charge on an electron and n is the principal quantum number.

From the above assumptions the gravitational binding energy is given as,

$$E_g = 2\left(\frac{\hbar c}{8\pi G}\right)^{3/2} \left(\frac{nc}{\alpha_e m_H}\right)^2 \qquad (25)$$

where α_e is the fine structure constant $\frac{ke^2}{\hbar c} = 1/137$

In Table 1 we list the values of E_g for several values of n-the principal quantum number. From this table it follows in particular, that the higher the principal quantum number, the higher the gravitational binding energy of a star.

The total gravitational binding energy of a star

n(Principal quantum number)	E_g (Joules)	Remarks
0.003212	3.229×10^{47}	
0.0345	3.468×10^{48}	White Dwarf
1	1.005×10^{50}	

What do we conclude from the foregoing calculation? We conclude that equation (25) is at the base of the equilbrium of actual stars in relation to the energy state and binding energy of the Hydrogen atom. It differs from the Chandrasker calculation by the introduction of a natural fine structure constant, providing the energy of proper magnitude for the measurement of stellar energies and therefore proving to be a better theory for stellar structure. This could be elaborated in detail by flowers original words,

"The Black-dwarf material is best likened to a single gigantic molecule in its lowest quantum state. On the Fermi-Dirac statistics, its high density can be achieved in one and only one way, in virtue of a correspondingly great energy content. But this energy can no more be expended in radiation than the energy of a normal atom or molecule. The only difference between Black-dwarf matter and a normal molecule is that the molecule can exist in afree state while the black dwarf matter can only so exist under high external pressure.

The Theory of White -Dwarf Stars and Black Holes; The Limiting Mass at the Lowest Principal Quantum Number

The gravitational energy is known to be of order $E_g = \frac{GM^2}{r}$, M being the mass of a star. Then equating this to equation (25) we obtain the radius of a star as,

$$r = \frac{1}{2}\left(\frac{8\pi G}{\hbar c}\right)^{3/2} \left(\frac{\alpha_e m_H}{nc}\right)^2 GM^2 \quad (26)$$

while the above equation states that the radius is proportional to the square of it's mass, the Chandrasker analysis is in disagreement, stating that r is inversely proportional to the cube root of the mass.

But at a point where r equation (26) approaches the schwarzichilds radius r_s

$$r \Rightarrow r_s, \frac{1}{2}\left(\frac{8\pi G}{\hbar c}\right)^{3/2} \left(\frac{\alpha_e m_H}{nc}\right)^2 GM^2 = \frac{2GM}{c^2}$$

We obtain an upper limit to the mass of,

$$M = 4\left(\frac{n}{\alpha_e}\right)^2 \left(\frac{\hbar c}{8\pi G}\right)^{3/2} \frac{1}{M_H^2} \quad (27)$$

Now consider equating the original solution of Chandrasker mass limit to our newly developed formula (27), we have

$$M_C = M(4), \quad \frac{\omega^0_3 \sqrt{3\pi}}{2}\left(\frac{\hbar c}{8\pi G}\right)^{3/2}\frac{1}{\mu_e^2 M_H^2}$$

$$= 4\left(\frac{n}{\alpha_e}\right)^2 \left(\frac{\hbar c}{8\pi G}\right)^{3/2} \frac{1}{M_H^2}$$

$$n = \frac{\alpha_e}{\mu_e}\left(\frac{\omega^0_3}{8}\sqrt{3\pi}\right)^{1/2}$$

$\omega^0_3 = 2.018236$, is a constant connected with the solution to the lane-Emden equation, and $\mu_e = 2$, average molecular weight per electron, then

$$n = 3.212 \times 10^{-3}$$

In the table below we list the values of M and r for several values of n-the principal quantum number, including the one calculated above.

The Mass limit and radius limit of a star

n(Principal quantum number)	M (Kilograms)	r (meters)	Remarks
3.212×10^{-3}	2.304×10^{28}	34.153	($0.012 M_{sun}$)
0.0345	2.66×10^{30}	3944.601	Chandrasekar mass limit ($1.4 M_{sun}$)
1	2.234×10^{33}	3.31×10^{6}	Maximum mass of a white dwarf

What do we conclude from the foregoing calculation? We conclude that the formation of a white dwarf star or any other stellar structure will never exceed the Schwarzichild's radius of 34.153m, this will only happen at the most lowest quantum principal number of 3.212×10^{-3}. For example, at the principal quantum number the size of the fine structure constant $\frac{1}{137}$, the mass obtained will be of $0.063 M_{sun}$ and r=176.443m. Therefore under high external pressure the minimum mass of a last star that is formed is of order $2.304 \times 10^{28} kg$ and this only occurs at r=34.153m under the lowest energy state below the known Bohrs radius of $a_o = 0.53 Å$.

What is Wrong With Hawking Temperature

In his paper "Particle creation by Black holes" Hawking pointed out that "In the classical theory black holes can only absorb and not emit particles. However it is shown that quantum mechanical effects cause black holes to create and emit particles as if they were hot bodies with temperature $\frac{\hbar c^3}{8\pi GMk} \approx 10^{-6} \left(\frac{M_{sun}}{M}\right)^o K$". However this is not the case when the assumptions given in the first sections of this book are taken into account. For example, we know that, the electric potential energy is given by,

$$E_e = \left(\frac{\hbar}{8\pi G}\right)^{3/2} \frac{c^{11/2}}{E_g m_H}$$

But treating the particles in the process General relativisticaly (at the Schwarzichild radius), the gravitational potential energy will be of order $E_g = \frac{mc^2}{2}$, giving the electric energy as,

$$E_e = 2\left(\frac{\hbar}{8\pi G}\right)^{3/2} \frac{c^{7/2}}{mm_H}$$

Now the thermal energy is given by $E_{thermal} = kT$, where k is the Boltzmann constant.

By the principal of Equipartition

$$E_{thermal} \sim E_e \Rightarrow T = 2\left(\frac{\hbar}{8\pi G}\right)^{\frac{3}{2}} \frac{c^{\frac{7}{2}}}{km\, m_H}, \qquad (28)$$

$$T = 3.3891 \times 10^{11} \frac{M_{sun}}{M}$$

Note: in a limit where m_H is the planck mass $m_H = \sqrt{\frac{\hbar c}{8\pi G}}$, equation 28 above for the temperature of a black hole reduces to the hawking temperature formula $T = \frac{\hbar c^3}{8\pi GMk}$

For conditions at the centre of the Sun, $T = 3.3891 \times 10^{11} K$ which is in disagreement with the Hawking temperature of $T_H = 6.476 \times 10^{-8} K$. This is left for the reader to analyse.

Entropy of a Black Hole

For derivations which i will not show here, I am led to the total energy of a Black Hole given by,

$$E_B = \frac{2A(\hbar c^{13})^{1/2}}{(8\pi G)^{5/2} m_H m}$$

where, A is the surface area of the event horizon

But since the entropy is energy per unit temperature,

$$S = \frac{E_B}{T}$$

Remember that temperature is given by equation(28),

Then the entropy will be given by,

$$S = \frac{Ac^3 k}{4\pi G \hbar}$$

This is in agreement with the Bekenstein-Hawking area entopy law

On the Development of a Quantum Gravity-Hydrostatic Equation and its Implication to Physics-Minimum Black hole mass

It is known that the equation governing the hydrostatic equilibrium of a star is given by,

$$\frac{dP}{dr} = -\frac{GM(r)}{r^2}\rho \qquad (29)$$

Where P denotes the total pressure, ρ is density, and M(r) is the mass interior to a sphere of radius r.

what if we rewrite the above formula in a form given by,

$$\frac{F_g F_e}{\hbar c} r^2 = \frac{c^4}{8\pi G} = constant \qquad (30)$$

Where F_g is the gravitational force, F_e is the electric force, c is the speed of light, $\hbar$ is the reduced Planck constant and G is the gravitational constant. Let the pressure be, $P = \frac{F_g F_e}{\hbar c}$, this means that pressure is dependent on the product of the gravitational and electric forces in a quantum-relativistic realm. Therefore in simple terms, we can write (30) in its simplest form as, $\frac{dP}{dr} = -\frac{c^4}{8\pi G r^2}(r)$ and to include the density, we have

$$\frac{dP}{dr} = -\frac{rc^2}{8\pi G r_s}(r)\rho$$

where r_s is schwarzichild's radius. thus at $r_s = r$, the star will form a black hole.

To differ from (29) we have formulated one of the first quantum gravity -hydrostatic equation. From (30) we can write the electric potential energy as,

$$F_e r = \frac{\hbar c^5}{8\pi G E_g}$$

where, $E_g = F_g r$ is the gravitational potential energy

at a point where the potential gravitational energy is in equilibrium with the potential electric energy the total energy is that of the Planck energy by,

$$E = F_e r = E_g = \sqrt{\frac{\hbar c^5}{8\pi G}} = 3.91 \times 10^8 \text{J}$$

since the Bohr energy of an hydrogen atom is given by,

$$E_n = \frac{m k_e^2 e^4}{2n^2 \hbar^2}$$

then, using the principle of equipartition of energy

$$F_e r = E_n$$

we deduce, the gravitational potential energy as

$$E_g = \frac{2n^2 \hbar^3 c^5}{8\pi G m_e k_e^2 e^4}$$

this can be written in a simplest form as,

$$E_g = \frac{2n^2}{m_e}\left(\frac{J_p}{\alpha_e}\right)^2 \qquad (31)$$

where, $J_p = \sqrt{\frac{\hbar c^3}{8\pi G}}$ is the planck momentum 1.3035N.s

$\alpha_e = \frac{ke^2}{\hbar c}$ is the fine structure constant $\frac{1}{137}$

m_e is the mass of an electron $9.11 \times 10^{-31} kg$

Then the total gravitational energy is calculated to be,

$$E_g = 7.00 \times 10^{34} J(n^2)$$

For a thermal energy kT, we estimate a temperature of

$$T = \frac{7.00 \times 10^{34}}{k} = 5.07 \times 10^{57} K$$

We know that, the gravitational potential energy is given by, $\frac{GM^2}{r}$, and for

$$E_g = \frac{GM^2}{r} = \frac{2n^2}{m_e}\left(\frac{J_p}{\alpha_e}\right)^2$$

The radius mass relation can be written as,

$$r = \frac{GM^2 m_e}{2n^2}\left(\frac{\alpha_e}{J_p}\right)^2$$

$$r = 9.527 \times 10^{-46} \frac{M^2}{n^2}$$

For the solar mass $M = 1.9 \times 10^{30} kg$, $r \sim 3.44 \times 10^{15} m$ and if r is equal to the schwarzichilds radius $\frac{GM}{c^2}$, then

$$\frac{GM}{c^2} = \frac{GM^2 m_e}{2n^2}\left(\frac{\alpha_e}{J_p}\right)^2$$

The solar mass limit is given by

$$M = \frac{2n^2}{m_e c^2}\left(\frac{J_p}{\alpha_e}\right)^2 \sim 7.78 \times 10^{17} \text{kg}$$

But for $r = \frac{2\pi\hbar}{m_e c}$, compton wavelength then

$$\frac{2\pi\hbar}{m_e c} = \frac{GM^2 m_e}{2n^2}\left(\frac{\alpha_e}{J_p}\right)^2$$

From which mass reduces to,

$$M = \frac{2n\hbar^{1/2}\pi^{1/2}}{G^{1/2} m_e c}\frac{J_p}{\alpha_e} \sim 2.913 \times 10^{12} \text{kg}$$

This is the minimum mass of a Black hole

Murder of Germans Sacred Cow: Experimental Test of General Relativity

It has long been suspected that the deflection of light in the vicinity of the sun exceeds the general relativistic predicted value of 1.75". An example of this, is the Erwin Finlay Freundlich 1929 solar eclipse expedition which produced a value of 2.24" larger than the general relativistic value. It is expected that once the reason for the deviation in the deflection angle has been found, it will disprove Einstein's imaginations for the curvature of space time. Although research into this field is scarce, we have managed through theoretical means under the assumption of a modified spherically symmetric solution to the Einstein field equation to prove E.F. Freundlich right. It is theorized that the bending of light near the sun is a function of the strength of the force (coupling constant) near the sun and the increasing distance from the sun's surface (in terms of the Schwarzschild radius).

It's almost hundred years since Sir Arthur Eddington experimentally proved Einstein's general relativity theory right. Since then, there has never been any competing theory that would prove Einstein wrong save for Loop quantum gravity and string theory. The fact that starlight is bent at the surface of the gravitating body by a deflection angle of 1.75" imposes a bound on the theoretical justification of gravity. Calculating an angle below or above 1.75" will be an upheaval in the founding blocks of physics. Erwin Finlay Freundlich was one of those people who stood out of the ordinary in 1929 when he published results with a larger angle of deflection than Eddington's.

An account on Freundlich 1929 expedition has been clearly given in Robert J.Trumpler and Klaus Hentschel papers as stated below;

"Among the various expeditions sent out to observe the total solar eclipse of May 9, 1929, that of the Potsdam Observatory (Einstein Stiftung) seems to be the only one which obtained photographs suitable for determining the light deflection in the Sun's gravitational field. Two instruments were used, but so far only the results of the larger one, a 28-foot horizontal camera combined with a coelostat, have been published. The three observers, Freundlich, von Klüber, and von Brunn, claim that these observations (four plates containing from seventeen to eighteen star images each) lead to a value of 2.24" for the deflection of a light ray grazing the Sun's edge; a figure that deviates considerably from the results of the 1922 eclipse, and which is in contradiction to Einstein'sgeneralized theory of relativity".

The irreducible anomaly in the observations of the deflection of light by the sun has been known to exist since the birth of Einstein General relativity theory. For example, in a 1959 classical review by A.A.Mikhailov, it concludes that observations yield instead of a general relativistic prediction of 1.75arcsec at the limb of the sun the simple mean value of 2.03 ± 0.10 over the GR prediction (see the table below, from Mikhailov 1959 analysis)

The existence of a 2.24" deflection angle by Freundlich, Von Kluber and Von Brunn therefore implies a requirement for the modification of the general theory of relativity. Science has evolved in this simpler manner of

Year	Number of plates	stars	Limits of r	Scale 1"=	p.s. of one star	A as given by observer	new reduction	Σv^2 A	$1".75$	straight line
1919	7	7	2·0– 5·4	28μ	0"·15	1·98	2·07±0·09	360	690	402
1922	4	71	2·1–13·0	22	0·13	1·72	1·83±0·11	425	446	419
1929	2	18	1·5– 7·5	41	0·15	2·24	1·96±0·08	1971	2372	3236
1936	2	29	2·0– 7·2	29	0·27	2·70	2·68±0·37	1375	1534	1630
1947	1	51	3·3–10·2	30	0·24	2·01	2·20±0·38	612	618	630
1952	2	11	2·1– 8·9	30	0·15	1·70	1·43±0·18	7058	8693	4039
Weighted mean 1"·93±0"·05 p.e.					Simple mean 2·03±0·10					

modifications although there are some who cling to the old thoughts of "The earth is the center of the universe and Einstein is always right". I am not proving anyone wrong but I want you to believe that the general relativity theory that was put forward by Einstein is not the only 'there is' excellent description of the universe, there are other ways far better than GR as it was with the Newtonian Gravitational force replacement with a curvature of space time.

K HENTSCHEL

FREUNDLICH in front (center) of the horizontal camera for measuring the deflection of light in the sun's gravitational field during the solar eclipse in Sumatra, 1929; from FREUNDLICH, V. KL/,)BER & V. BRUNN [1931]b PlateII or [1931]a p. 176.

In this section I will prove Erwin Finlay Freundlich solar eclipse results right but from a theoretical perspective. We base our study on the bending of starlight past the surface of the sun, we establish the deflection angle at which this occurs starting from General relativity and beyond.

Einstein's theory proposes that gravity is not an actual force, but is instead a geometric distortion of spacetime not predicted by ordinary Newtonian physics. The more mass you have to produce the gravity in a body the more distortion you get, this distortion changes the trajectories of objects moving through space, and even the paths of light rays, as they pass close-by the massive body. Even so, this effect is very feeble for an object as massive as our own sun, so it takes enormous care to even detect that it is occurring.

General Relativity predicts how much of this bending of light you should see given the mass of the object. By formula, the Einstein General Relativity deflection angle is given by,

$$\theta_{GR} = \frac{4GM_\odot}{c^2 R_\odot} = \frac{2R_s}{R_\odot} = 1.75 \text{arcsec}$$

Where, $M_\odot$ — mass of Sun(1.989×10^{30} kg)

$R_\odot$ — Radius of Sun(6.957×10^8 m) and $R_s = \frac{2GM_\odot}{c^2}$ is the Schwarzischild radius. The above given value doesn't come by surprise and it is the work of genius to find out why Newton, Einstein and others got different values. Below we provide a solution to the anomally in the observation of light deflection at the sun's surface as given below.

For a test particle or an observer falling freely from infinity to a distance r_0 from the gravitating body, the modified spherically symmetric solution to the Einstein field equation (see Eqn 7 in chap1) will be given by;

$$(\theta R)^2 = \frac{1}{\alpha^{1/2}} \left(\frac{R_s r_0}{2} \right)$$

Where, $R_s = \frac{2GM}{c^2}$ is the Schwarzschild radius of a gravitating body, α is the coupling constant and θ is the angle of deflection of a light ray past a gravitating body. This angle was never introduced in the previous chapter but it has been introduced in here for purposes supporting this research. In what follows, we use the above equation by subsitituting in the values of the dimensionless physical constants for the theory in question to obtain values of the deflection angle for each theory. This analysis will help us recover new theories based on the coupling constant and then draw conclusions.

Let us start with the Newton's theory of gravitation. To recover the Newtonian deflection angle at the suns limb, $r_o = R_s$, we set the dimensionless physical constant to be $\alpha = 0.25$. This then gives the Newtonian value as,

$$\theta_1 = \frac{R_s}{R_\odot} = 0.875 \text{arcsec}$$

The Einstein General relativistic value can be got in the same way but this time with the dimensionless physical constant given by $\alpha = 0.0156$. This then gives the GR value as,

$$\theta_2 = \frac{2R_s}{R_\odot} = 1.75 \text{arcsec}$$

The Freundlich deflection angle might have taken a different twist than with Eddington 1.75arcsec result, which we are yet to find out and which is the reason for this expedition. Takin, $r_o = R_s$ and $\alpha = 5.8208 \times 10^{-3}$, we deduce the deflection angle given by the formula,

$$\theta_3 = \frac{2.56 R_s}{R_\odot} = 2.24 \text{arcsec}$$

Lastly when the dimensionless constant is the fine structure constant, $\alpha_e = \frac{1}{137}$ (note that, we have not used the gravitational coupling constant here simply because light as a photon is masssles). We get the following deflection angle,

$$\theta_4 = \frac{2.426 R_s}{R_\odot} = 2.12 \text{arcsec}$$

Our first result from the above calculations is that; the mean of the deflection angles from the four observations gives the exact deflection angle that was calculated and observed by Eddington in General relativity as,

$$\frac{\sum_{n=1}^{4} \theta_n}{4} = \frac{0.875 + 1.75 + 2.24 + 2.12}{4}$$
$$= 1.75 arcsec$$

The difference in the observed deflection angle is due to variations in the coupling constant or dimensionless physical constant from the modified metric which to a great extent differs from the Rindler, Schwarzschild, Reissner-Nordstrom, Kerr-Newman and Friedman-Lemaitre metrics. The fact that the mean of the four observations for the deflection of light given above reproduces the GR value of 1.75arcsec imposes a general bound on the dimensionless physical constants which is one of the unsolved problems in physics. Keeping other factors constant, the sum of four values of the coupling constants in any observation must not exceed the following value,

$$\alpha_1 + \alpha_2 + \alpha_3 + \alpha_4 = \frac{1}{256} = 3.90625 \times 10^{-3}$$

The model given above still assumes that general relativity is the correct theory of gravity on cosmological scales. If this was not correct, then the mean will have produced a different value of the deflection angle. But the fact that the mean reproduced the GR value makes it the correct theory of gravity from all the theories involved. Therefore our part is to determine the kind of theory behind the Freundlich deflection angle. And once this is found out then there will be transparence in the final theory, unified theory and the theory of everything.

What is Special about the Energy Density?

Physicists believe that gravity is not a real force like the electromagnetic or the strong force because an inertial frame of reference eliminates the effects of gravity. One way to prove that gravity is a true force is if a gravitational field can be shown to possess energy density. When two masses undergo unsymmetrical acceleration, they emit quadrupole gravitational waves that definetely possess energy. A similar acceleration of two of the same polarity charged particles produces quadrapole electromagnetic radiation. The question is, if the electric and magnetic field store energy, what about the gravitational field? The first insight into answering the question came from Albert Einstein in his own words he stated that;

"to the intra-atomic movement of electrons, atoms would have to radiate not only electromagnetic but also gravitational energy if only in tiny amounts, as this is hardly true in nature, it appears that quantum theory would have to modify not only Maxwellian electrodynamics, but also the new theory of gravitation".

Gravitational radiation is produced when massive bodies accelerate. This radiation is difficult to detect due to the weakness of the gravitational force. It can only be detected under vigorous observations of the radiations from supernovae and collisions of black holes. The study of the gravitational radiation would come straight from the Bohr's theory of an atom but it proves difficult since one cannot even deduce the intensity of the electromagnetic wave from such a theory. Rather the intensity of radiation emitted from an atom is studied using the known formula for the intensity in electromagnetism (EB /μo). To clearly understand the intensity of the electromagnetic wave, one needs to develop a formula for the intensity of a wave on a quantum scale. Once such is formulated, it would then become easy to perform calculations for the intensity of the gravitational waves.

To differ from Bohr's model of hydrogen atom, it is hereby theorized that; the total energy of an atom is related to the electric, magnetic and gravitational forces as,

$$W_p = \frac{mgEe}{Bev} r = \frac{F_G F_e}{F_B} r \qquad (32)$$

Where $F_G = mg$ is the gravitational force, $F_e = Ee$ is the electric force on an electron in vicinity of an electric field and $F_B = Bev$ is the magnetic force

To determine the strength of the electromagnetic force on a quantum scale, we borrow an analogy from the theory of quantum mechanics by which the quantized angular momentum is deduced from the fine structure constant and denoting the coupling constant to behave as the principle quantum number (Remember our aim here is to deduce the intensity of a wave emitted from an atom due to an electron performing Bohr's orbits), In formula we express the quantized angular momentum of an electron due to an electromagnetic interaction as,

$$\frac{Ke^2}{c} = n\hbar \qquad (33)$$

Where ℏ is the reduced plank constant, c is the speed of light, k is the coulomb constant (k=1 /4πε$_o$, ε$_o$ is permeability of free space) and n is the principle quantum number.

Since the gravitational force is almost negligible in an atom, it becomes a catastrophe to treat it as a quantum mechanical effect. In quantum mechanics the angular momentum of an electron is quantized in units of nℏ while in gravitational mechanics, there is no such thing as quantization, which is why we treat gravitation classically. Then the formula for the angular momentum due to the gravitational force will be given by,

$$\frac{Gm^2}{c} = mvr \qquad (34)$$

It is therefore evident that the gravitational descriptions of an electron can only be treated classically. This is why it has proven difficult to merge gravity with quantum mechanics. However such a problem has been solved here, by considering the assumptions below,

The speed of light in both quantum and gravitational processes is a constant and therefore if we substitute the speed of light from eqn (33) into eqn(34) we get the expression for the angular momentum as,

$$mvr = \frac{F_G}{F_e} n\hbar \qquad (35)$$

We have thus introduced the ratio of the gravitational force to the electric force in the formula for Bohr's quantized angular momentum. This ratio represents the negligible gravitational effects in an atom. It is therefore a correction to the Bohr's atomic model.

Because the gravitational force can be expressed in many ways by using Eqn(32), we can deduce the power carried by the electromagnetic wave due to the motion of an electron in an atom. Making F_G the subject from equation (32) and substituting for it in equation (35), we get the power as,

$$F_B c = \frac{2\pi r^2 \lambda m v F_e^2}{nh^2} \qquad (36)$$

Since the de Brogile wave length is $\lambda = \frac{h}{mv}$, and the surface area of the sphere is $A = 4\pi r^2$. Then the Intensity of a wave from a particle exhibiting both wave and particle properties is

$$\frac{F_B c}{A} = \frac{F_e^2}{2nh} = \frac{E^2 e^2}{2nh} \qquad (37)$$

Keeping other factors constant we have theorized that, the intensity of a wave is proportional to the square of the electric field, a fact that would be impossible to deduce in Bohr's atomic model. The above formula can only be deduced if only we take into account (in theory), the combined effects of gravity and electromagnetism.

On the other hand, If we let the power of the electromagnetic wave be $P = F_B c$, and n be the fine structure constant $\alpha = ke^2/\hbar c$, then the equation for the intensity of the electromagnetic wave comes out clearly as, $P = EB/\mu_0 = 2\varepsilon_0 E^2 c$, Where μ_0 is the permeability of free space.

The problem of finding the energy density stored in the gravitational field reflects the incompleteness of general relativity. The problem of the free falling clocks which have to maintain the same rate is something related to the problem of non accountable gravitational energy. Feyman tried to put a patch on the issue of the gravitational energy but wasn't very successful either.

The problem is of paramount importance since it is reasonably something which impairs the unification of quantum theory and gravity. Newtonian gravitation is conservative globally and locally, GRT is not, quantum theory is based strictly on energy and momentum conservation.

A precise and consistent quantum theory of gravity has not yet been proved, not even by the self proclaimed geniuses of this time. We are aware and satisfied that classical General Relativity is the most precise description of gravity due to its predictable nature. The left hand side of Einstein field equation represents the metric of space time curvature while the right hand side represents the matter - energy content of the classical matter fields of pressure and energy density. It is known that quantum mechanics plays an important role in the behaviour of the matter fields but has no place in the Einsteins field equations.

In its simplest form the Einstein Field equation relates the cosmological constant Λ to the energy density ρ as,

$$\Lambda = \frac{8\pi G}{c^4} \rho$$

Where, G is the gravitational constant, and c is the constant speed of light.

The problem of reconciling the quantum theory with general relativity is brought about by not knowing the quantum mechanical version of the energy density, we only know the classical energy density in the electromagnetic field but the quantum mechanical energy density from which the electromagnetic energy density can be derived is not known. This is a very big problem faced by researchers in the field of quantum gravity. This wasn't only a problem to A. Einstein but also to Stephen.W.Hawking. In his paper "Particle creation by Black holes" Hawking wrote about the problem in this way, "one therefore has a problem of defining a consistent scheme in which the space time metric is treated classically but is coupled to the matter fields which are treated quantum mechanically"

By creating a correct quantum mechanical energy density from which both the electromagnetic and gravitational energy density can be derived we will be able to solve the quantum gravity problem. But before we do, let me present to you a quick insight into the Larmor power law which led to all of the problems we are facing today. Failure of the Bohr model and all quantum mechanics models to give a correct solution to the larmor problem is what is limiting us from finding the quantum theory of gravity.

For example, the laws of classical mechanics (i.e. the Larmor formula),

$$P = \frac{q^2 a^2}{6\pi\varepsilon_0 c^3}$$

Where a is the proper acceleration, q is the charge, and c is the speed of light.

predict that the electron will release electromagnetic radiation while orbiting a nucleus. Because the electron would lose energy, it would rapidly spiral inwards, collapsing into the nucleus on a timescale of around 16 picoseconds. This atom model is disastrous, because it predicts that all atoms are unstable. Also, as the electron spirals inward, the emission would rapidly increase in frequency as the orbit got smaller and faster. This would produce a continuous smear, in frequency, of electromagnetic radiation. However, late 19th century experiments with electric discharges have shown that atoms will only emit light (that is, electromagnetic radiation) at certain discrete frequencies. To overcome this hard difficulty, Niels Bohr proposed, in 1913, what is now called the *Bohr model of the atom*. He put forward these three postulates that sum up most of the model:

1. The electron is able to revolve in certain stable orbits around the nucleus without radiating any energy contrary to what classical electromagnetism suggests. These stable orbits are called stationary orbits and are attained at certain discrete distances from the nucleus. The electron cannot have any other orbit in between the discrete ones.

2. The stationary orbits are attained at distances for which the angular momentum of the revolving electron is an integral multiple of the reduced Planck's constant: , where n = 1, 2, 3, ... is called the principal quantum number, and $\hbar = h/2\pi$. The lowest value of n is 1; this gives a smallest possible orbital radius of 0.0529 nm known as the Bohr radius. Once an electron is in this lowest orbit, it can get no closer to the proton. Starting from the angular momentum quantum rule, Bohr was able to calculate the energies of the allowed orbits of the hydrogen atom and other hydrogenlike atoms and ions. These orbits are associated with definite energies and are also called energy shells or energy levels. In these orbits, the electron's acceleration does not result in radiation and energy loss. The Bohr model of an atom was based upon Planck's quantum theory of radiation.

3. Electrons can only gain and lose energy by jumping from one allowed orbit to another, absorbing or emitting electromagnetic radiation with a frequency v determined by the energy difference of the levels.

To overcome all the problems arising from the Larmor power formula we developed a new formula which relates the energy density to the force (and which is a generalized field formula from Equation37) it is called the quantum mechanics energy density,

$$\rho = \frac{F^2}{8\pi\alpha\hbar c}$$

Where ρ is the energy density stored in the field of the force F, α is the coupling constant that determines the strength of the force, and $\hbar$ is the reduced Planck constant.

Therefore the larmor power formula that was given above can be modified to be,

$$P = \frac{AF^2}{8\pi\alpha\hbar} = \frac{Am^2a^2}{8\pi\alpha\hbar}$$

Where, A is the surface area of orbit of a particle emiting the radiations around the nucleus of an atom, m is the mass of the particle, and a, is the proper acceleration.

The above given power formula will reduce to the Larmor power formula only when the area A is limited by,

$$A = \frac{4}{3\varepsilon_0 c^3}\left(\frac{q}{m}\right)^2 \alpha\hbar$$

Which implies that the area swept out by an electron in orbit around the nucleus of an atom is quantized. This is a classical version of loop quantum gravity in which the area of space occupied by a particle is quantized.

From the new larmor formula given we can clearly see that the charge does not appear in the formula. This then takes us to our original problem of determining the energy density in the gravitational field.

We know that the electric field store energy, and that in a vacuum the energy density is given by, $\rho = \frac{\varepsilon_0}{2} E^2$ where E is the electric Field and ε_0 the permittivity of free space. If our new formula for the enegy density given above is true, it must be able to reproduce the expression for the energy density of the electric field and also solve other problems.

To derive the energy density in the electric field, we let the force on the particle say an electron with charge e due to the electric field E created by another charged electron be, F=eE. Then the energy density will be related to the electric field by,

$$\rho = \frac{e^2 E^2}{8\pi \alpha \hbar c}$$

But because the coupling constant of the electromagnetic force is the fine structure constant $\alpha = \frac{e^2}{4\pi \varepsilon_0 \hbar c}$, then on substitution and cancelling like terms, we recover the energy density in the electric field as,

$$\rho = \frac{\varepsilon_0}{2} E^2$$

Similary, for the energy density in the gravitational field, let the force experienced by a particle of mass m due to the gravitational field g be F=mg. The energy density is here given by,

$$\rho = \frac{m^2 g^2}{8\pi\alpha\hbar c}$$

But because the coupling constant of the gravitational force is the fine structure constant- $\alpha = \frac{Gm^2}{\hbar c}$, then on substitution and cancelling like terms, we recover the energy density in the gravitational field as,

$$\rho = \frac{g^2}{8\pi G}$$

We have shown that, just as the electromagnetic field stores energy, the same is also true for the gravitational field.

Then the Einstein field equation can be summarized in the following format,

$$\Lambda = \frac{8\pi G}{c^4}\rho = \frac{8\pi G}{c^4}\left(\frac{F^2}{8\pi\alpha\hbar c}\right)$$

$$\Lambda = \frac{GF^2}{\alpha\hbar c^5} = \frac{F^2}{\alpha E_{pl}^{\;2}}$$

Where $E_{pl} = \sqrt{\frac{\hbar c^5}{G}}$ is the Planck energy in Planck units.

We have therefore proved that the space time metric which is treated classically can be coupled to the matter fields which are treated quantum mechanically by the introduction of the energy density that is treated quantum mechanically.

Therefore in a limit where $F = \frac{c^4}{G}$ (Newtonian limit), we have the cosmological constant given by,

$$\Lambda = \frac{c^3}{\alpha\hbar G} = \frac{1}{\alpha l_p^{\;2}}$$

Where $l_p = \sqrt{\frac{\hbar G}{c^3}}$ is the Planck length.

We have shown that the unification of quantum mechanics with general relativity implies that there is a fundamental length in Nature in the sense that no operational procedure would be able to measure distances shorter than the Planck length. Finally, using hand waving arguments we have also shown that a minimal length might be related to the cosmological constant which, if this scenario is realized, is time dependent.

The coupling constant is now related to the force between particles and to the cosmological constant by the following formula,

$$\alpha = \frac{GF^2}{\Lambda \hbar c^5} = \frac{F^2}{\Lambda E_{pl}^2}$$

Such that when the force between two particles is the gravitational force $= \frac{GMm}{R^2}$, and $\Lambda = \frac{R_s^2}{4R^4}$, ($R_s$ is Schwarzchild radius) we get the usual known gravitational coupling constant, $\alpha = \frac{GMm}{\hbar c} = \frac{E_g^2}{E_{pl}^2}$ (E_g is

Schwarzchild Energy). This gives a simplest way of calculating the coupling constant.

Application of the quantum energy density to space time singularities, information paradox, Planck stars, Emergence of the laws of Newton, galaxy rotation problem and the Tully-Fisher relationship

We consider the possibility that the energy of a collapsing star and any additional energy falling into the Black hole could condense into a highly compressed core with density of the order of the Planck density. If we let the quantum force pressing on the surface of a star be given as,

$$F_q = \frac{\hbar c}{R^2}$$

Where R is the radius of a Black hole, ℏ is the reduced Planck constant and c is the constant speed of light.

Let also the gravitational attraction force opposing the quantum force from within the collapsing star be

$$F_b = \frac{c^4}{G\alpha}$$

Where G is the gravitational constant and α is the coupling constant

The quantum energy density is given as,

$$\rho = \frac{F^2}{8\pi\alpha\hbar c}$$

Subsitituting for F_b in the above given energy density we have,

$$\rho = \frac{c^7}{8\pi\alpha^3\hbar G^2}$$

Therefore nature appears to enter the quantum gravity regime when the energy density of matter reaches the Planck scale. The point is that this may happen well before relevant lengths become planckian. For instance, a collapsing spatially compact universe bounces back into an expanding one. The bounce is due to a quantum-gravitational repulsion which originates from the modified Heisenberg uncertainty, and is akin to the force that keeps an electron from falling into the nucleus. Therefore bounce does not happen when the universe is of planckian size, as was previously expected; it happens when the matter energy density reaches the Planck density. At this energy density, a Planck star is formed. The key feature of this theoretical object is that this repulsion arises from the energy density, not the Planck length, and starts taking effect far earlier than might be expected. This repulsive 'force' is strong enough to stop the collapse of the star well before a singularity is formed, and indeed, well before the Planck scale for distance. Since a Planck star is calculated to be considerably larger than the Planck scale for distance, this means there is adequate room for all the information captured inside of a black hole to be encoded in the star, thus avoiding information loss.

If this is the case, the gravitational collapse of a star does not lead to a singularity but to one additional phase in the life of a star: a quantum gravitational phase where the gravitational attraction is balanced by a quantum pressure and that is, when $F_q = F_b$

$$R = \alpha^{1/2} \left(\frac{\hbar G}{c^3}\right)^{1/2}$$

$$R = \alpha^{1/2} l_p$$

Where $l_p = \left(\frac{\hbar G}{c^3}\right)^{1/2}$ is the Planck length

For instance, if n = 1/3, and α is the gravitational coupling constant, a stellar-mass black hole would collapse to a Planck star with a size of the order of 10^{-10} centimeters. This is very small compared to the original star in fact, smaller than the atomic scale but it is still more than 30 orders of magnitude larger than the Planck length. This is the scale on which we are focusing here. The main hypothesis here is that a star so compressed would not satisfy the classical Einstein equations anymore, even if huge compared to the Planck scale. Because its energy density is already planckian.

Derivation of Newton's law of gravity

Modified Newtonian dynamics (MOND) is a theory that proposes a modification of Newton's laws to account for observed properties of galaxies. It is an alternative to the theory of dark matter in terms of explaining why galaxies do not appear to obey the currently understood laws of physics. By applying our quantum mechanical energy density we explain why gravity is emergent and why the velocities of stars in galaxies were observed to be larger than expected based on Newtonian mechanics.

From the energy density relationship, the new effective gravitational force is related to quantum energy density by,

$$F = (8\pi\alpha\hbar c\rho)^{1/2}$$

Since energy density is the amount of energy stored in a given system or region of space per unit volume, let a body of mass M store an amount of energy $E = Mc^2$ (due Einstein mass energy relationship) in a given region per unit volume $V = \frac{4}{3}\pi R^3$ where R is the radius. The energy density of this volume of space will be given by,

$$\rho = \frac{3Mc^2}{4\pi R^3}$$

If the dimensionless coupling number is given by,

$$\alpha = \alpha_g \frac{R_s}{12R}$$

Where $R_s = \frac{2GM}{c^2}$ is the Schwarzichild radius of a body of mass M, and $\alpha_g = \frac{Gm^2}{\hbar c}$ is the gravitational coupling constant of a particle of mass m.

From the above given expression, the coupling number increases with an increase in the Schwarzchild radius but decreases with an increase in the radius R. For a particle m at a distance R so great from M the coupling number is very small.

When the expression of the energy density and coupling number are substituted into our force formula above, we get the usual Newton's law of gravitation as,

$$F = \left(8\pi\alpha_g \frac{R_s}{12R} \hbar c \left(\frac{3Mc^2}{4\pi R^3}\right)\right)^{1/2}$$

$$F = \frac{GMm}{R^2}$$

From the above given derivation it has been shown that a particle m feels a force (i.e gravity) due to the energy density of mass M. The mass m comes from the gravitational coupling constant. This simply shows that gravity originates from the vacuum energy density and distance originates from the coupling number.

Alternative Proof of Newton's Law of Universal Gravitation

Previously we showed that the energy density ω is related to the force F by,

$$\omega = \frac{F^2}{8\pi\alpha\hbar c}$$

Where, h is the reduced Planck constant, c is the constant speed of light and is the coupling constant or principal quantum number

In this short notice we clearly prove that F is the gravitational force that was put forward by Newton as we are yet to see below,

Independent of the mass and distance, the force between two particles as in the case of the Casmir affect is therefore given by,

$$F = \sqrt{8\pi\alpha\hbar c\omega} \qquad (38)$$

Below I give two conditions on which the above force will reduce to the Newton's universal law of gravitation

The energy density is related to the cosmological constant by,

$$\omega = \frac{m^2 c^3}{8\pi \hbar} \Lambda \qquad 39$$

This was previously derived and m was the Planck mass

Here m is taken to represent the mass of the particle in circular orbit around a massive body of mass M at a distance or radius of curvature R from M. Where, $\Lambda = \frac{1}{R^2}$.

The coupling constant or principal quantum number is here given as a ratio of the areas as,

$$\alpha = \frac{A_s}{A} \qquad 40$$

Where $A_s = \frac{4\pi G^2 M^2}{c^4}$, is the Schwarzschild area occupied by a massive body and $A = 4\pi R^2$ is the total area of circular orbit of a mass m.

Substituting condition 39 and 40 into equation (38) above we obtain the Newton's law of gravitation as,

$$F = \frac{GMm}{R^2}$$

This derivation is proof that gravity is a result of the quantum vacuum energy density. While this has proved to be a short insight into the emergency of gravity, we shall have a long discussion of this research in the coming chapters.

The Tully-Fisher Relation

While Newton's laws predict that stellar rotation velocities should decrease wih distance from the galactic centre, Rubin and collaborators found instead that they remain almost constant. The rotation curves are said to be flat. This observation necessitates either one of the following, 1) there exists in galaxies large quantities of unseen matter which boosts the stars velocities beyond what would be expected on the basis of the visible mass alone, or 2) Newton's laws do not apply to galaxies. The former leads to the dark matter hypothesis; the latter leads to Modified Newtonian dynamics (MOND)

Newton's laws works well in high acceleration environments, that is in the solar system and on Earth while it fails for objects with extremely low acceleration, such as stars in the outer parts of galaxies. To resolve the problem we have proposed a new effective gravitational force law given by,

$$F = (8\pi\alpha\hbar c\rho)^{1/2}$$

Applying this to an object of mass m in circular orbit around a point mass M (a crude approximation for a star in the outer regions of a galaxy), we find:

$$\frac{mv^2}{R_1} = (8\pi\alpha\hbar c\rho)^{1/2}$$

Since energy density is the amount of energy stored in a given system or region of space per unit volume, let a body of mass M store an amount of energy $E = Mc^2$ (due Einstein mass energy relationship) in a given region per unit volume V^*. The energy density of this volume of space will be given by,

$$\rho = \frac{Mc^2}{V^*}$$

Since the gravitational coupling constant for mass m is known to be,

$$\alpha = \frac{Gm^2}{\hbar c}$$

On substitution and cancelling like terms, we have

$$v^4 = \frac{8\pi c^2 R_1^2}{V^*} GM$$

Where v is the star's rotation velocity at a distance R_1 from the center of the galaxy, the rotation curve becomes flat, as required only when the following relation is correct,

$$\frac{V^*}{R_1^2} = 1.885 \times 10^{28} m = R_o$$

Which gives

$$v^4 = \frac{8\pi c^2}{R_o} GM = a_o GM$$

That is, the star's rotation velocity is independent of R_1, its distance from the centre of the galaxy- the rotation curve is flat, as required. By fitting this law to rotation curve data, we have found the Milgrom acceleration $a_o = 1.2 \times 10^{-10} m/s^2$.

For a sun's orbit around our galaxy, the radius of the sun's orbit is $R_1 = 8000pc$, ($1pc = 3.0857 \times 10^{16} m$) this implies a volume of the space bound by matter at the centre of our milky way galaxy to be, $V^* = 1.149 \times 10^{69} m^3$.

This simple law is sufficient to make predictions for a broad range of galactic phenomena

How to Calculate a Mysterious Repulsive Force Pulling Galaxies Apart

It is known that in a homogenous cosmological universe, a positive cosmological constant induces repulsive forces. The question; is there a classical formula of the force of the cosmological constant like that of the gravitational force? How does the repulsive force relate to the cosmological constant and the coupling constant? How does understanding the energy density in relation to force, change the way we perceive Einstein's field equation? The section sets out to answer these and more questions about the cosmological constant problem. (see chapter 26 on the origin of gravity and electrodynamics)

Dr Lee Smolin represents the perimeter institute for theoretical physics. He claims that the mathematisation of physics has resulted in the reduction of the cosmos to a mathematical entity, which has not only confused physicists but accounts for their worst and most distracting assertions.

There is a wide spread speculation that the mathematical formulation of physics has not only confused physicists but has also lead to failures in the development of a quantum theory of gravity.

Although both general relativity and quantum mechanics work well in the domain of their applicability, it's unfortunate that there is no unified theory of gravity with quantum mechanics.

It is proposed that the unification of gravity with quantum mechanics will require us to change the kind of mathematics that was used by either Einstein or Schrödinger etal.. in the development of both theories. But why do we bother at all if there is another way in which we can express the theory better without the use of tensor fields.

The problem with the mathematical formulation of general relativity if at all it exists stems from the non existence of its experimental observation which wasn't the case with quantum mechanics. The formulation of quantum mechanics was based on the existence of experimental observations. Therefore quantum mechanics was founded on the existence of experiments which wasn't the case with general relativity. Einstein had to base his theorization on thought experiments which could or wouldn't be nearer to any experimental confirmation of the phenomenon being studied.

The same is also true for the formulation of quantum gravity. There is no sound experimental proof for the existence of quantum gravitational effects and therefore scientists like Hawking Stephen have also clung to the old formulations that were used by Einstein and his contemporaries to develop a quantum theory of gravity.

In this brief notice we show that an existence of a unified theory is rooted deep into the unnoticed pressure-energy density similar to the stress energy tensor appearing in Einstein field theory. Our major aim therefore is to provide proof for the questions set out below;

(i) If the cosmological constant introduces a force of repulsion between bodies. Is it true that the force increases in simple proportion to the cosmological constant and the coupling constant?

(ii) Is there a classical formula of the force of the cosmological constant like that of the gravitational force?

Einstein's general relativity equations famously described the curvature of space-time as the mechanism for gravity. In the original theory, Einstein added a "cosmological constant" that acted as an expulsive force to counteract gravity. That stabilized the universe so it didn't collapse in on itself, but Einstein abandoned the idea when further astronomical observations showed the universe was accelerating and not static, as the great physicist had thought.

Analogous to the known Einstein field equation, the curvature of space (cosmological constant) is here related to the energy density ω as,

$$\Lambda = \kappa\omega = \kappa\left(\frac{F^2}{8\pi\alpha\hbar c}\right) \quad (41)$$

Where $\kappa = \frac{8\pi G}{c^4}$ a constant appearing in Einstein's field equation, F is is the force in an interaction and α is the coupling constant.

The above expression implies that the cosmological constant is related to the force and therefore increases as a square of the force.

For the energy density in electric field, where $F = Ee$ and $\alpha = \frac{e^2}{4\pi\varepsilon\hbar c}$, the energy density will be given by, $\omega = \frac{F^2}{8\pi\eta\hbar c} = \frac{\varepsilon E^2}{2}$.

While for the energy density in the gravitational field, where $F = mg$ and $\alpha = \frac{Gm^2}{\hbar c}$, the energy density will be given by, $\omega = \frac{F^2}{8\pi\eta\hbar c} = \frac{g^2}{8\pi G}$. This can be written in simple terms as $\omega = \frac{\eta g^2}{2}$, where $\eta = \frac{1}{4\pi G}$.

From (41) therefore, the force responsible for the expansion of the universe is related to the cosmological constant by,

$$F = E_{pl}(\alpha\Lambda)^{1/2} \quad (42)$$

Where $E_{pl} = 1.9605 \times 10^9 J$ is the Planck energy.

Given the Planck (2015) values of $\Omega_\Lambda = 0.6911\pm0.0062$ and $H_o = 67.74\pm0.46$ (km/s)/Mpc $= (2.195\pm0.015)\times10^{-18}$ s^{-1}, Λ has the value of 1.11×10^{-52}m^{-2} as given in wikimedia commons.

Based on the above given value, the force will then have a value of

$$F_{ob} = 2.0655 \times 10^{-17}(\alpha)^{1/2}$$

$$F_{ob} \sim 1.8 \times 10^{-18} N$$

This therefore is a force responsible for the expansion of the Universe. It is such a small force that will require sophiscated machines to measure. While the above force value is based on the fine structure constant, there is a value that is even smaller than that value by, $F_{ob} \sim 1.58 \times 10^{-36}$N at $\alpha = 5.87 \times 10^{-39}$ between two protons.

However in quantum electrodynamics (QED) we compute a much larger value of $F_{QED} \sim 2.82 \times 10^{44}(\alpha)^{1/2}$ N. This huge discrepancy is known as the cosmological constant problem. Therefore the relative strength of the force will be given by;

$$\frac{F_{QED}}{F_{ob}} \sim 10^{61}$$

The above value is in agreement with the Hubble age to the Planck time, which is the same as the total mass of the universe to the Planck mass as,

$$\frac{F_{QED}}{F_{ob}} = \frac{t_H}{t_{pl}} = \frac{M_U}{M_{pl}} \sim 10^{61}$$

The above given relationship implies a persistence constant error that is evident when comparing observational and theoretical calculations. This error needs to be distributed uniformly in order to correct for large discrepancies which accrue to calculated values in relation to observed values.

The problem lies in knowing the observed force value to the calculated value, since the force ratio doesn't correspond to the other ratios of time and mass. In other words changing the ratio $\frac{F_{QED}}{F_{ob}}$ to $\frac{F_{ob}}{F_{QED}}$ will cause other ratios to change.

It is therefore observed that the ratio of Hubble age to the Planck time and the total mass of the universe to the Planck mass will only be in line or tally with the Planck force to the Hubble force by a value $\sim 10^{61}$ and not otherwise.

Keeping other factors constant it is clear from the above given observations that the mysterious, repulsive force pulling galaxies apart is proportional to the coupling constant value in a given interaction. This proposal will be of such a great importance to the work of researchers involved in the field of quantum gravity

A Simple Link between MONDian Dynamics and the Dark Universe

According to H.Sabine (2018), in the disc galaxies most of the mass is at the centre of the galaxy, this means that if you want to calculate how a star moves far away from the centre it is a good approximation to only ask what is the gravitational pull that comes from the centre bulge of the galaxy. Einstein taught us that gravity is really due to the curvature of space and time but in many cases it is still quantitatively incorrect to describe gravity as a force, this is known as the Newtonian limit and is a good approximation as long as the pull of gravity is weak and objects move much slower than the speed of light. It is a bad approximation for example close by the horizon of a black hole but it is a good approximation for the dynamics of galaxies that we are looking at here. It is then not difficult to calculate the stable orbit of a star far away from the centre of a disc galaxy. For a star to remain on its orbit, the gravitational pull must be balanced by the centrifugal force, $\frac{mv^2}{R} = \frac{GMm}{R^2}$. You can solve this equation for the velocity of the star and this will give you the velocity that is necessary for a star to remain on a stable orbit, $v = \sqrt{\frac{GM}{R}}$. As you can see the velocity drops inversely with the square root of the distance to the centre. But this is not what we observe, what we observe instead is that the velocity continue to increase with distance from the galactic centre and then they become constant.

This is known as the flat rotation curve. This is not only the case for our own galaxy but it is the case for hundred of galaxies that have been observed. The curves don't always become perfectly constant sometimes they have rigorous lines but it is abundantly clear that these observations cannot be explained by the normal matter only.

Dark matter solves this problem by postulating that there is additional mass in galaxies distributed in a spherical halo. This has the effect of speeding up the stars because the gravitational pull is now stronger due to the mass from the dark matter halo. There is always a distribution of dark matter that will reproduce whatever velocity curve we observe.

In contrast to this, Modified Newtonian Dynamics (MOND) postulates that gravity works differently. In MOND, the gravitational potential is the logarithmic of the distance $\Phi = \left(\sqrt{GMa_o}\right)\ln\left(\frac{R}{GM}\right)$, and not as normally the inverse of the distance $\Phi = \frac{-GM}{R}$.

In MOND the gravitational force is then the derivation of the potential that is, the inverse of the distance $F = \frac{\sqrt{GMa_o}}{R}$, while normally it is the inverse of the square of the distance $F = \frac{GMm}{R^2}$. If you put this modified gravitational force into the force balance equation as before $\frac{\sqrt{GMa_0}}{R} = \frac{v^2}{R}$, you will see that the dependence on the distance cancels out and the velocity just becomes constant. Now of course you cannot just go and throw out the normal $\frac{1}{R^2}$ gravitational force law because we know that it works on the solar system. Therefore MOND postulates that the normal $\frac{1}{R^2}$ law crosses over into a $\frac{1}{R}$ law. This crossover happens not at a certain distance but it happens at a certain acceleration. The New force law comes into play at low acceleration a_0, this acceleration where the crossover happens is a free parameter in MOND. You can determine the value of this pararmeter by just trying out which fits the data best. It turns out that the best fit value is closely related to the cosmological constant $a_0 \approx \sqrt{\frac{\Lambda}{3}}$, why does that so? No one has any idea and it is the aim of this section to find out why.

The cosmological constant is a specific type of dark energy but as for now there is no known relation between dark energy and dark matter and it is the work of this paper to find out.

A formula for apparent dark matter energy density in galaxies and clusters

Attempts to find a way to measure or to calculate the value of the vacuum energy density and the cosmological constant have all either failed or produced results incompatible with observations or other confirmed theoretical results. Some of those results are theoretically implausible because of certain unrealistic assumptions on which the calculation model is based. And some theoretical results are in conflict with observations, the conflict itself being caused by certain questionable hypotheses on which the theory is based. And the best experimental (Casmir effect) evidence is based on the measurement of the difference of energy density within and outside of the measuring apparatus, thus preventing in principle any numerical assessment of the actual energy density. Below we present a formula for the energy density that accurately calculates the vacuum energy density and which is in agreement with observations.

$$\rho = \frac{F^2}{8\pi \alpha_e \hbar c} \qquad (43)$$

Where, F is Force, α_e is the coupling, $\hbar = h/2\pi$ is the reduced Planck constant and c is a constant speed of light

Now if we consider dark matter to be a Planck particle with a Planck mass $M_{pl} = \sqrt{\frac{\hbar c}{G}}$ under the influence of a gravitational force $F = M_{pl} a_o$, where a_o is the low gravitational acceleration in MOND.

Hence, when we put $F = M_{pl} a_o$ (which is the gravitational pull on a Planck particle of Dark matter in a low acceleration due to gravity of MOND) in the formula (1) we obtain a relation between the energy density and the MOND low acceleration a_o.

$$\rho = \frac{M_{pl}^2 a_o^2}{8\pi \alpha_e \hbar c} \qquad (44)$$

This is the main formula and central result of our paper, since it allows one to make a direct comparison with observations.

As a final fun comment let us, just out of curiosity, take the formula (44) and apply it to the entire universe. Now we note that the critical energy density of the universe equals

$$\rho_{crit} = \frac{3 H_o^2 c^2}{8\pi G} = \frac{M_{pl}^2 a_o^2}{8\pi \alpha_e \hbar c}$$

Since $M_{pl} = \sqrt{\frac{\hbar c}{G}}$ we obtain a relation between a_o (MONDian- determined by fits to internal properties of galaxies), H_o is the hubble constant (a measure of the present-day expansion rate of the Universe) and α_e (coupling which determines the strength of the force) as,

$$a_o = cH_o\sqrt{3\alpha_e} \qquad (45)$$

This relation holds remarkably well for the experimentally verified parameters and because it is in agreement with observations, our energy density formula (44) would be applicable to the entire universe.

A formula for the cosmological constant

The cosmological constant Λ is a dimensionful parameter with units of $(length)^{-2}$. From the point of view of classical general relativity, there is no preferred choice for what the length scale defined by Λ might be. Particle physics, however, brings a different perspective to the question. Einstein introduced a cosmological constant into his equations for General Relativity. This term acts to counteract the gravitational pull of matter, and so it has been described as an anti-gravity effect. The cosmological constant turns out to be a measure of the energy density but no one has ever calculated the cosmological constant with confidence. Below we create a formula for the cosmological constant that is related to a_o and which is in agreement with observations.

$$\Lambda = \frac{a_o^2}{\alpha_e c^4} \qquad (46)$$

Hence, when we put $a_o = cH_o\sqrt{3\alpha_e}$ (45) in our formula (46) we obtain a relation between the Hubble constant and the cosmological constant

$$H_o = c\sqrt{\frac{\Lambda}{3}} \qquad (47)$$

But when the constant speed of light c in formula (47) $c = H_o\sqrt{\frac{3}{\Lambda}}$ is put into our formula (46) we then have

$$a_o = 3H_o^2\sqrt{\frac{\alpha_e}{\Lambda}} \qquad (48)$$

The relation above clearly shows that a_o is also close to the acceleration rate of the universe H_o, and hence the cosmological constant Λ.

Finally when we put the value of $a_o{}^2 = \Lambda \alpha_e c^4$ from formula (4) into our formula for the energy density (44) we then recover an approximation to the Einstein Field equation of General relativity as,

$$\rho = \frac{M_{pl}{}^2 c^3}{8\pi \hbar} \Lambda \tag{49}$$

Since $M_{pl} = \sqrt{\frac{\hbar c}{G}}$ we obtain a relation between Λ (cosmological constant) and ρ (vacuum energy density) as was put forward by Einstein in his General relativity theory.

$$\Lambda = \frac{8\pi G}{c^4} \rho \tag{50}$$

Therefore from general formulas and assumptions given above, we have constructed a full theory that connects a_o (MOND acceleration), H_o (Hubble constant), Λ (Cosmological constant), α_e (Fine structure constant) and c (constant speed of light) in a natural way.

Determination of the Hubble constant, Cosmological constant and the Vacuum Energy Density

In this section we determine the values of the vacuum energy density (ρ), Hubble constant (H_o) and Cosmological constant (Λ) without invoking dark matter. This happens when the parameters a_o (determined by fits to internal properties of galaxies) and α_d (the coupling constant) take on values $a_o = 1.2 \times 10^{-10} m/s^2$ and $\alpha_d = \frac{2}{124.89} \approx \frac{2}{125}$ within galaxies and galaxy clusters. The results we obtain are in agreement with the Planck collaboration parameters of 2018. What remains is to confirm by experiment the value of the coupling constant α_d and once confirmed, the equations given will prove once and for all that the postulated Dark matter hypothesis is not responsible for what happens in galaxies and galaxy clusters.

The values of the vacuum energy density and cosmological constant have been observed experimentally by Planck collaboration (2018) to be $5.364 \times 10^{-10} J/m^3$ and $1.11 \times 10^{-52} m^{-2}$ respectively but there is no promising theory which can determine these values with confidence. Attempts to find a way to measure or to calculate the value of the vacuum energy density and the cosmological constant have all either failed or produced results incompatible with observations or other confirmed theoretical results. Some of those results are theoretically implausible because of certain unrealistic assumptions on which the calculation model is based. And some theoretical results are in conflict with observations, the conflict itself being caused by certain questionable hypotheses on which the theory is based. And the best experimental (Casmir effect) evidence is based on the measurement of the difference of energy density within and outside of the measuring apparatus, thus preventing in principle any numerical assessment of the actual energy density. To precisely determine the energy density (ρ) and the cosmological constant Λ, we create laws which work well at all field strength levels (i.e the electromagnetic field, the gravitational field, etc). The formula for the cosmological constant is known from the Friedmann solution to the Einstein field equations

$$\Lambda = \frac{8\pi G}{c^4} \rho \qquad (51)$$

Where, $c = 299792458 m/s$ is the speed of light and $G = 6.67430 \times 10^{-11} m^3 kg^{-1} s^{-2}$ is the gravitational constant.

Since no one has ever accurately calculated the value of the vacuum energy density that is in agreement with observations, I'm therefore tempted to be the first person to come up with a formula for the vacuum energy density which appears in (51) above. It is hereby theorized that the energy density is proportional to the square of the force field F as,

$$\rho = \frac{F^2}{8\pi\alpha\hbar c} \qquad (52)$$

Where, α is the coupling constant, $\hbar = \frac{h}{2\pi} = 1.054571817 \times 10^{-34} J.s$ is the reduced Planck constant.

We know that our formula (52) accurately calculates the energy density stored in the electric field E in this way; if we consider an electron in the electric field, the electric force felt by this electron due to the intensity of the electric field is given as, F=Ee, where e is the elementary charge on an electron. Putting F into (2) we obtain the energy density stored in the electric field as

$$\rho = \frac{E^2 e^2}{8\pi\alpha_e\hbar c} \qquad (53)$$

But because the coupling of the electromagnetic force is known to be, $\alpha_e = \frac{e^2}{4\pi\varepsilon_0\hbar c} = 1/137.036$, on putting this into our formula(3), the energy density stored in the electric field is

$$\rho = \frac{\varepsilon_0 E^2}{2} \qquad (54)$$

That is what we exactly deduce from the electromagnetic field theory (Maxwell electromagnetic theory).

When we apply our formula (52) to galaxies and galaxy clusters we entirely get something different. Unlike in electromagnetism where an electron in the electric field feels an electric force F=Ee at a coupling sacle $\alpha_e = 1/137.036$, in galaxies we assume that a particle with a Planck mass $M_{pl} = 2.18 \times 10^{-8} kg$, will feel a force $F = M_{pl} a_o$ due to the gravitational field strength $a_o = 1.2 \times 10^{-10} m/s^2$ at the coupling scale of $\alpha_d = 2/124.89 \approx 2/125$ (this coupling constant is weaker than the electromagnetic coupling by 0.46 because we are no longer dealing with electrons). Comparing this coupling to the formula for the gravitational coupling constant, $\alpha_G = \frac{GM^2}{\hbar c} = 2/124.89$, we obtain the mass for this coupling to be $M = \sqrt{\frac{2\hbar c}{124.89 G}} = 2.76 \times 10^{-9} kg$. This observation my have implications for a varying gravitational constant of $G_v = \frac{124.89}{2} G = 4.165 \times 10^{-9} Nm^2/kg^2$. This has not been confirmed experimentally but it points to the assumption that dark matter is made up of Planck mass relics (or primordial black holes) which remain whenever a black hole evaporates. The parameter a_0 is the acceleration scale introduced in the phenomenological fitting formula for galaxy rotation curves. a_o is the parameter that was introduced by Milogram in MOND. It is also an explanation for the phenomenological success of Milogram's fitting formula, in particular in reproducing the flattening of rotation curves, where the asymptotic velocity of the flattened galaxy rotation curve is, $v_f^4 = a_o GM$ this is known as the baryonic Tully-Fisher relation and has been well tested by observations of a very large number of spiral galaxies.

Hence, when we put $F = M_{pl} a_o = 2.616 \times 10^{-18} kg$ in the formula (52) we obtain a relation between the energy density and the acceleration a_o

$$\rho = \frac{M_{pl}^2 a_o^2}{8\pi \alpha_d \hbar c} \tag{55}$$

This is the main formula and central result of our sect, since it allows one to make a direct comparison with observations. When we put the Planck mass $M_{pl} = \sqrt{\frac{\hbar c}{G}}$ into our formula (55) above we find that the Planck mass disappears from our formula without changing our results and we remain with α_d to be determined. Remember we had earlier postulated that $\alpha_d = 2/124.89$ (which remains to be determined experimentally). Putting all those assumptions into consideration we obtain the formula for the vacuum energy density as

$$\rho = \frac{a_o^2}{8\pi G \alpha_d} = 5.369 \times 10^{-10} J/m^3 \tag{56}$$

Putting our formula (56) into the cosmological constant formula (51) we precisely determine the value of the cosmological constant to be

$$\Lambda = \frac{a_o^2}{\alpha_d c^4} = 1.11 \times 10^{-52} m^{-2} \qquad (57)$$

As a final fun comment let us, just out of curiosity, take the formula (56) and apply it to the entire universe. Now we note that the critical energy density of the universe (from Friedmann solution) equals

$$\rho_{crit} = \frac{3H_o^2 c^2}{8\pi G} = \frac{a_o^2}{8\pi \alpha_d G} \qquad (58)$$

But since the density parameter, Ω, is defined as the ratio of the actual (or observed) density ρ to the critical density ρ_{crit} of the Friedmann universe we then have

$$\Omega = \frac{\rho}{\rho_{crit}} = \frac{a_o^2}{3H_o^2 c^2 \alpha_d} \qquad (59)$$

We then obtain a relation between a_o (MONDian-determined by fits to internal properties of galaxies) and H_o -the Hubble constant (a measure of the present-day expansion rate of the Universe) as

$$a_o = cH_o\sqrt{3\Omega\alpha_d} \tag{60}$$

Putting a_o into our formula for the cosmological constant (7) we obtain the Hubble constant as

$$H_o = c\sqrt{\frac{\Lambda}{3\Omega}} = \frac{1.82483 \times 10^{-18} s^{-1}}{\sqrt{\Omega}} \tag{61}$$

The value of has been determined experimentally to be, $\Omega = 0.6889 \pm 0.056$ Planck(2018), therefore the Hubble constant is precisely determined to be $H_o = 2.1986 \times 10^{-18} s^{-1}$. This relation holds remarkably well for the experimentally verified parameters and because it is in agreement with observations, our energy density formula (52) would be applicable to the entire universe.

Therefore from general formulas and assumptions given above, we have provided a precise calculation of the cosmological parameters of ρ (vacuum energy density), Λ (Cosmological constant) and H_o (Hubble constant) in agreement with the Planck (2018) observations. We have assumed only one parameter α_d which we believe will be determined by experiment in future. The assumed Planck mass may not necessary be assumed as this disappears from our equations and therefore does not affect the result. The results of ρ and Λ only happen when $a_o = 1.2 \times 10^{-10} m/s^2$ and $\alpha_d \approx \frac{2}{125} = 0.016$. Because the parameter a_0 has been determined to be a best fit for galaxy rotation curves, we therefore remain to determine by experimental means the coupling constant α_d for galaxies and galaxy clusters. Once α_d is confirmed, the equations given will prove once and for all that the postulated Dark matter hypothesis is not responsible for what happens in galaxies and galaxy clusters.

Derivation of the Temperature and Entropy of Black Holes

If the semi-diameter of a sphere of the same density as the sun were to exceed that of the sun in the proportion of 500 to 1, a body falling from an infinite height towards it would have acquired at its surface greater velocity than that of light, and consequently supposing light to be attracted by the same force in proportion to its vis inertiae, with other bodies, all light emitted from such a body would be made to return towards it by its own proper gravity (John Michell).

The development of general relativity followed a publication of acceleration under special relativity in 1907 by Albert Einstein. In his article, he argued that any mass will "Distort" the region of space around it so that all freely moving objects will follow the same curved paths curving toward the mass producing the distortions. Then in 1916, Schwarzschild found a solution to the Einstein field equations, laying the groundwork for the description of gravitational collapse and eventually black holes.

By definition, a black hole is an astronomical object with a very strong gravitational effect, which disturbs particles across its event horizon. It is also true from the theory of general relativity, that even light cannot escape its gravitational pull. These objects have puzzled the minds of great thinkers for many years. History puts it that, they were first predicated by John Michell and Pierre-Simon Laplace in the 18th century but David Finkelstein was the first person to publish a promising explanation of them in 1958 based on Karl Schwarz child's formulations of a solution to general relativity that characterized black holes in 1916.

In 1971, Hawking developed what is known as the second law of black hole mechanics in which the total area of the event horizons of any collection of classical black holes can never decrease, even if they collide and merge. This is similar to the second law of thermodynamics which states that, the entropy of a system can never decrease.

In 1972 Bekenstein proposed an analogy between black hole physics and thermodynamics in which he derived a relation between the entropy of black hole entropy and the area of its event horizon.

$$S = \frac{Akc^3}{4G\hbar}$$

In 1974, Hawking predicted an entirely astonishing phenomenon about black holes, in which he ascertained with accuracy that black holes do radiate or emit particles in a perfect black body spectrum.

$$T = \frac{\hbar c^3}{8\pi GMk}$$

Hawking beautiful result raises a number of questions. First, in Hawking's derivation the quantum properties of gravity are neglected. Are these going to affect the result? Second, we understand macroscopical entropy in statistical mechanical terms as an effect of the microscopical degrees of freedom. What are the microscopical degrees of freedom responsible for the entropy? Can we derive the entropy from first principles?

This book presents a new approach to Black hole thermodynamics that is different from that given by Loop Quantum Gravity (LQG), String theory and Bekenstein-Hawking radiation theory. The major result of the book is the derivation of the Bekenstein-Hawking area entropy law from first principles using new methods with a well defined calculation where no infinities appear. As far as this book is concerned there is no other theory from which such a calculation can proceed. Hence the methods used in here are the only one from which a detailed quantum theory of gravity precedes and where the result of the Bekenstein-Hawking area entropy law can be achieved.

Proof of the bekenstein-hawking black hole entropy law from first principles

From our energy density equation, since power is energy W per unit time t, we have the amount of energy added to the black hole as;

$$W = \frac{AtF^2}{8\pi\alpha\hbar}$$

Let the gravitational force of a black hole be $F = \frac{c^4}{8\pi G}$ and the time taken by the black hole to dissipate or evaporate be $t = \frac{16\pi^2\alpha\hbar}{MC^2}$, where MC^2 is the Einstein Energy of a Black hole. Then on substitution into the above given formula we have the energy as,

$$W = \frac{Ac^6}{32\pi G^2 M}$$

This is the Frodden-Ghosh- Perez Energy. From the Clausius definition of entropy, the entropy S of a black hole is the quantity of heat or energy W added to a Black hole per unit temperature of a black hole T.

$$S = \frac{W}{T}$$

From Hawking original calculation, the temperature of a black hole is given by,

$$T = \frac{\hbar c^3}{8\pi GMk}$$

Therefore the Entropy of a Black hole will be given by,

$$S = \frac{\left(\frac{Ac^6}{32\pi G^2 M}\right)}{\left(\frac{\hbar c^3}{8\pi GMk}\right)} = \frac{Akc^3}{4G\hbar}$$

Black hole Volume Entropy law

Still from the new power formula, the energy density is energy per unit volume V given as,

$$\rho = \frac{W}{V} = \frac{F^2}{8\pi\alpha\hbar c}$$

Let the gravitational force of a black hole be $F = \frac{c^4}{8\pi G}$, then the energy added to the black hole is related to the Volume of the black hole by;

$$W = \frac{Vc^7}{512\pi^3 \alpha \hbar G^2}$$

From the Clausius definition of entropy, the entropy S of a black hole is the quantity of heat or energy W added to a Black hole per unit temperature of a black hole T.

$$S = \frac{W}{T}$$

From Hawking original calculation, the temperature of a black hole is given by,

$$T = \frac{\hbar c^3}{8\pi GMk}$$

Therefore the Entropy of a Black hole will be given by,

$$S = \frac{\left(\frac{Vc^7}{512\pi^3 \alpha \hbar G^2}\right)}{\left(\frac{\hbar c^3}{8\pi GMk}\right)} = \frac{VkMc^4}{64\pi^2 \alpha \hbar^2 G}$$

Therefore the entropy of a black hole is related to the Volume of the black hole by the above given formula.

The formula given above, reduces to the area entropy law formula earlier given only when the volume of the black hole is,

$$V = \frac{16\pi^2 \alpha \hbar A}{Mc} = 16\pi^2 \alpha \lambda_c A$$

Where $\lambda_c = \frac{\hbar}{Mc}$ is the Compton wavelength of a Black hole and A is its area.

In conclusion, the article has presented a new approach to Quantum mechanics that is different from that given by Bohr, Heisenberg, Schrödinger and others. The major result of the research is the derivation of the Bekenstein-Hawking area entropy law from first principles using new methods with a well defined calculation where no infinities appear. As far as this book is concerned there is no other theory from which such a calculation can proceed. Hence the methods used in here are the only one from which a detailed quantum theory of gravity "Holy Grail of modern physics" precedes and where the result of the Bekenstein-Hawking area entropy law can be achieved.

Particle Creation by Black Holes: Is it Hawking's Approach or My Approach?

Can we ever hope to find the right way? Nay more, has this right way an existence outside our illusions? I will answer without hesitation that there is, in my opinion, a right way, and that we are capable of finding it. Our experience hitherto justifies us in believing that nature is the realization of the simplest conceivable mathematical ideas. I am convinced that we can discover by means of purely mathematical constructions the concepts and the laws connecting them with each other, which furnish the key to the understanding of natural phenomena.

Albert Einstein (1879-1955)

In 1975 Hawking calculated quantum mechanically that a black hole will emit particles as if it were a black body with a temperature proportional to its surface gravity. Although this thermal emission is insignificant for black holes formed by stellar collapse, it is of crucial importance for the small primordial black holes formed by density fluctuations in the early universe.

The most significant consequence of a black hole is that, the temperature of a black hole increases as a black hole loses mass. The temperature increases exponentially into a burst of gamma rays leaving a black hole remnant.

There is no clear account on this, not until we have fully developed a consistent quantum theory of gravity (where the mass of a black hole approaches the Planck scale of mass and radius). The evaporation of a black hole starts with a spin down phase in which the Hawking radiation carries away the angular momentum, after which it proceeds with emission of thermally distributed quanta until the black hole reaches the Planck mass.

The radiation spectrum contains all Standard Model particles, which are emitted on our brane, as well as gravitons, which are also emitted into the extra dimensions. It is expected that most of the initial energy is emitted during this phase in Standard Model particles.

One of the major problem with black holes is that, we cannot directly measure any properties of them neither can we produce black holes in any terrestrial experiment. According to Cheung (2002), this is due to the fact that in order to produce black holes in collider experiments one needs a centre of mass energy above the Planck scale, which is obviously inaccessible at the moment. But thanks to the introduction of the numerical coefficient, we can now as this book directs, detect a vast number of black holes in our galaxy by observing and detecting the mass scale or the low energy scale quanta emitted whenever a black hole evaporates due to stellar collapse.

The numerical coefficient α depends on which particle species can be emitted at a significant rate and can be determined by taking the effect of the absorption cross section. This coefficient is of great importance in the standard model and if described in detail it could unlock the secrets hidden deep in the cosmos.

The dominant contribution to α in the standard model comes from fermions, the contribution to α for electrons and positrons is 1.575×10^{-4} (Don Page 1975). Page calculated the emission rates for massless particles, predicted the lifetime of black holes (from the total power emitted in all modes) and also deduced the numerical coefficient for the dimensionally determined quantities (in terms of the Planck mass etc). The coefficient appears in Eqn25 and Eqn26 of the rate of change of mass and the life time of a black hole by Don Page (1975). There is no known formula relating the numerical coefficient to the mass scale or low energy scale quanta emitted the mass of an electron and the mass of a proton. Yet this could provide a unique probe of at least four areas of physics: the early Universe; gravitational collapse; high energy physics; and quantum gravity.

Assuming that a black hole emits particles at a mass scale M_* (low energy scale quanta), we propose the Numerical coefficient to be

$$\alpha = \frac{2M_* m_e}{m_p^2} \qquad (62)$$

Where, m_p is the mass of the Proton and m_e is the mass of an electron. This concept will help us understand the type of particles emitted by Black holes and how detecting them will help us observe most of the Black holes in space, the coeffient will be efficient in the derivation of the Chandrasekher mass limit and the Bekenstein-Hawking area entropy law. Also this coefficient will be of good use in understanding the lowest energy level of the White dwarf Hydrogen atom. In the table below, we give values of α for different mass scales M_*.

Table 1

α	Mass scale M_* (kg)	Remarks
2.837×10^{15}	4.343×10^{-9}	Planck particle-Planck scale
1	1.531×10^{-24}	Yet to be found $\approx$ 1TeV
0.073	2.23×10^{-25}	Higgs boson
1.575×10^{-4}	2.41×10^{-28}	Pion-Neutral- cosmic rays

Note: M_{pl} (the Planck mass of $\left(\frac{\hbar c}{8\pi G}\right)^{1/2}$ =4.343 × 10^{-9}kg)

From the analysis given above, a black hole of mass M_{BH} will have a temperature and a life time given by

Temperature: $T = \dfrac{M_{pl}^2 c^2}{k M_{BH}} \left(\dfrac{2 M_* m_e}{m_p^2}\right) = \dfrac{M_{pl}^2 c^2}{k M_{BH}} \alpha$

$$T = 8.0305 \times 10^{46} \dfrac{M_*}{M_{BH}}$$

Lifetime: $\tau = \dfrac{G M_{BH}^3 m_p^2}{M_{pl}^2 c^3 M_* m_e} = \dfrac{2 G M_{BH}^3}{M_{pl}^2 c^3 \alpha}$

$$\tau = 8.019 \times 10^{-43} \dfrac{M_{BH}^3}{M_*}$$

Note: The power of a black hole is given by, $P = \dfrac{M_{pl}^2 c^5}{2 G M_{BH}} \alpha$

For purposes of this study, let us limit ourselves to two Primordial Black holes, one with a mass of 4.7×10^{11}kg and another with mass 1.331×10^{17}kg. We calculate the Temperature and life time of these black holes at known and assumed mass scales as given in table 2 and table 3.

Table 2.

Black hole (Kg)	Mass scale M_*	Temp- T(K)	τ (sec)	Remarks
4.7 × 10¹¹	4.343×10^{-9}	7.42×10^{26}	19.17	Early universe
	1.531×10^{-24}	2.62×10^{11}	5.44×10^{16}	Current Age of the Universe

2.23 × 10⁻²⁵	3.81 × 10¹⁰	3.73 × 10¹⁷	Current Age of the Universe
2.41 × 10⁻²⁸	4.12 × 10⁷	3.46 × 10²⁰	?

Note: For $M_* = M_{BH}$ we obtain the temperature of a Black hole $T = 8.0305 \times 10^{46}$ K. This is the maximum temperature of a black hole above which the black hole cease to exist.

Table 3.

Black hole (Kg)	Mass scale M_*	Temp- T(K)	τ (sec)	Remarks
1.33 × 10¹⁷	4.343 × 10⁻⁹	2.62 × 10²¹	4.34 × 10¹⁷	Current Age of the Universe
	1.53 × 10⁻²⁴	9.24 × 10⁵	1.23 × 10³³	

2.23×10^{-25}	1.35×10^5	8.46×10^{33}
2.41×10^{-28}	145.52	7.83×10^{36}

We learn from the above tables that, the temperature and lifetime associated with a black hole will not only depend on the mass of a black hole but also on the mass scale of the quanta emitted as scaled from the numerical coefficient which depends on which particle species can be emitted at a significant rate. For example, the theory of black hole radiations that was developed by S.W. Hawking will only become correct and deducible to the Hawking temperature and life time formula for black holes in a limit $M_* = 1.531 \times 10^{-24}$kg and $\alpha = 1$. Such that, $T = \frac{1.229 \times 10^{23} \text{kg}}{M_{BH}}$ $^{\circ}$K and $t = 5.238 \times 10^{-19} M_{BH}^3$. In other words M_* is assumed to be the scale of the underlying theory. The predictions of the Hawking radiations for a black hole with mass 4.7×10^{11}kg are as given in table 2 at a mass scale 1.531×10^{-24}kg. These take on similar properties for the Higgs boson. Therefore observations at such a scale could shed more light on the detection of a 4.7×10^{11}kg black hole. If we observe at a scale of a Pion $M_* = 2.41 \times 10^{-28}$kg at the current age of the universe (about 13.8×10^9yrs) we should be able to detect a black hole with a mass of 6.698×10^4kg (Primordial Black hole).

The ideas presented above could provide a unique probe of at least four areas of physics: the early Universe; gravitational collapse; high energy physics; and quantum gravity. The first topic is relevant because studying primordial black hole formation and evaporation can impose important constraints on primordial inhomogeneities and cosmological phase transitions. The second topic relates to recent developments in the study of "critical phenomena" and the issue of whether primordial black holes are viable dark matter candidates. The third topic arises because primordial black hole evaporations could contribute to cosmic rays, whose energy distribution would then give significant information about the high energy physics involved in the final explosive phase of black hole evaporation. The fourth topic arises because it has been suggested that quantum gravity effects could appear at the TeV scale ($M_* = 1.531 \times 10^{-24}$ kg) and this leads to the intriguing possibility that small black holes could be generated in accelerators experiments or cosmic ray events, with striking observational consequences (see B.J.Carr, 2005).

Lastly, the significance of α-the numerical coefficient can be seen in a broad sense when applied to the sun. If we take the sun to be a black hole with mass 1.99×10^{30} kg and a temperature at its center of $T = 1.5 \times 10^7$ K, we obtain a mass scale of $M_* = 3.717 \times 10^{-10}$ kg, which gives a life time of $\tau = 1.700 \times 10^{58}$ sec, the time that will be taken by the sun to dissipate if the temperature given at its center was 1.5×10^7 K.

The Chandrasekhar Mass Limit

A region in the universe has a potential energy of self-gravitation,

$$E_g = \frac{M_{pl}^2 c^2}{M_{BH}} \left(\frac{2M_* m_e}{m_p^2}\right)\left(\frac{6.144\pi^3}{\mu_e^2}\right) \quad (63)$$

A star will collapse to a White dwarf when the above energy is in equilibrium with the energy due to the electron degeneracy pressure of a Hydrogen atom given as,

$$E_e = m_e c^2$$

Where $\mu_e = 2$ is the average molecular weight per electron, which depends upon the chemical composition of a star. Then for, $E_g = E_e$

$$M = \frac{12.288\pi^3}{\mu_e^2} \frac{M_{pl}^2 M_*}{m_p^2}$$

In a limit for $M_* = M_{pl}$ and $\alpha = 2.837 \times 10^{15}$, we obtain the Chandrasekhar mass limit for a white dwarf star as,

$$M = \frac{12.288\pi^3}{\mu_e^2} \frac{M_{pl}^3}{m_p^2} = 1.4 M_{sun}$$

Note that such a result is only possible in the given limit but for a limit such as $M_* = 1.531 \times 10^{-24}$ kg we obtain, $M = 4.956 \times 10^{-16} M_{sun}$ which is the mass of the Primordial black hole, providing evidence for the Hawking limit for particle emission by black holes as described previously

Planck epoch

From an expression for the life time of a black hole, it is theorized that a Black hole has a mass $M_{BH} = kM_*^{1/3}$ where k is a constant. For k=1, we have a life time of 8.019×10^{-43} sec almost the Planck time-the earliest period of time in the history of the universe).

The Bekenstein-Hawking area entropy law

From the Black hole temperature we can calculate the entropy of a black hole, the total energy of a black hole with mass M and surface area A is given as,

$$E = \frac{Ac^5 M_{pl}{}^2 M_* m_e}{2\pi G\hbar m_p{}^2 M}$$

The change in entropy when a quantity of E is added to a black hole is,

$$S = \frac{E}{T}$$

Since the temperature is known (see above) on substituting we have

$$S = \frac{Ac^3 k}{4\pi G\hbar}$$

This is the Bekenstein-Hawking area entropy formula.

The Lowest Possible Energy State of a White dwarf Hydrogen Atom

In this section we prove an existence of the minimal principal quantum number which imposes a general bound on the energy level of the Hydrogen atom and the orbital radius of an electron. The results are derived from general laws not known by the entire scientific community. The section therefore provides a relationship between the micro and macro structures of the universe at a level when the atomic mass limit is in equal proportion to the Chandrasekhar mass limit.

I won't go into details of the literature of the Chandrasekhar mass limit as these have been repeatdly written and analysed in almost a million papers about the topic. But for a brief introduction into the derivation of the Chandrasekhar mass I refer the reader to Chandrasekhar 1983 Noble prize lecture (1). Almost every aspect of a white dwarf star has been studied but there is one thing which we do not know about white dwarfs in relation to the Hydrogen atom and this is encoded in Flower's original statement;

"The black-dwarf material is best likened to a single gigantic molecule in its lowest quantum state. On the Fermi-Dirac statistics, its high density can be achieved in one and only one way, in virtue of correspondingly great energy content. But this energy can no more be expended in radiation than the energy of a normal atom or molecule. The only difference between black-dwarf matter and a normal molecule is that the molecule can exist in a free state while the black-dwarf matter can only so exist under very high external pressure"

The question is, do we have an existing relationship between the mass limit of the Hydrogen atom and the White dwarf star? If the black-dwarf material is best likened to a single gigantic molecule in its lowest quantum state, what is the lowest possible energy state at which such a relationship exists?

Briefly let us propose in formula a model to support our argument; Firstly, let the potential energy of self-gravitation of a star be given by,

$$E_g = \frac{2M_{pl}{}^3 m_e c^2}{M_s m_{pro}{}^2}\left(\frac{6.144\pi^3}{\mu_e{}^2}\right) \qquad (64)$$

Where M_{pl} is the Planck mass $\left(\frac{\hbar c}{8\pi G}\right)^{1/2}$, m_{pro} is the Proton mass, m_e electron mass, M_s mass of star and c is the constant speed of light

Lastly, the quantized energy of an Hydrogen atom is given by,

$$E_n = \frac{m_e K_e{}^2 e^4}{2n^2 \hbar^2} \qquad (65)$$

Where n is the principle quantum number which indicates the energy levels in the Hydrogen atom.

By connecting the above equations we shall be able to deduce the lowest principle quantum number in the Hydrogen atom, providing one of the first relationship between the microscope and macroscopic structures of the universe.

Equating (64) to (65) we have

$$\frac{2M_{pl}^3 m_e c^2}{M_S m_{pro}^2}\left(\frac{6.144\pi^3}{\mu_e^2}\right) = \frac{m_e K_e^2 e^4}{2n^2 \hbar^2}$$

On arranging and canceling like terms we have;

$$n = \frac{\alpha_e \mu_e m_{pro}}{2}\sqrt{\frac{M_S}{6.144\pi^3 M_{pl}^3}}$$

Where $\alpha_e = \frac{K_e e^2}{\hbar c} = \frac{1}{137}$ is the fine is is structure constant.

When, $M_s = 1.4 M_{sun}$ (Chandrasekhar mass limit), we have $n = 5.1586 \times 10^{-3}$. This is the allowed principal quantum number or the lowest energy state of an Hydrogen atom for a white dwarf star at the Chandrasekhar mass limit. Therefore the quantized energy of the Hydrogen atom at this principal number is

$$E_n = \frac{13.606 eV}{n^2} = 511.289 \times 10^3 eV$$

And the electron radius at this energy level is $r = 1.405 \times 10^{-15} m$.

This result implies that, whereas the Bohr's orbital quantization doesn't permit orbits below the Bohr radius of $5.28 \times 10^{-11} m$, the theory above says that this is possible for an atom under high pressure. The electrons are therefore bound to the surface of the proton. For a white dwarf star of $M_s = 0.87 M_{sun}$, we have $n = 4.07668 \times 10^{-3}$. This gives the radius of a proton of $r = 8.775 \times 10^{-16} m$ which has been determined by spectroscopy methods

Is There A Limit To How Small Black Holes Can Become?

The smallest black hole would be one where the Schwarzschild radius equals the radius of a mass with a reduced Compton wavelength which is the smallest size to which a given mass can be localized. For a small mass M, the Compton wavelength exceeds half the Schwarzschild radius, and no black hole description exists. This smallest mass for a black hole is thus approximately the Planck mass, the micro black hole.

Contrary to the above observation, torsion (see Einstein-Cartan theory) modifies the Dirac equation in the presence of the gravitational field causing fermions to be spatially extended. This spatial extension of fermions limits the minimum mass of a black hole to be on the order of 10^{16} Kg, showing that micro black holes (of Planck mass) may not exist. Another mass limit is from the data of the Fermi Gamma-ray space telescope satellite which states that, less than one percent of dark matter could be made of primordial black holes with masses up to 10^{13} Kg.

The major aim of this section is to prove theoretically the existence of a minimum mass limit of a black hole and thereafter prove Chandrasekhar wrong (see Chandrasekhar 1983 Noble lecture concluding statement below)

"We conclude that there can be no surprises in the evolution of stars of mass less than 0.43Solarmass ($\mu = 2$). The end stage in the evolution of such stars can only be that of the white dwarfs. (Parenthetically, we may note here that the so-called 'mini' black-holes of mass 10^{12} Kg cannot naturally be formed in the present astronomical universe.)"

From the theory of white dwarf stars, the radius limit of a white dwarf of mass M is given by the following equation,

$$R_w = \frac{(9\pi)^{2/3}}{8} \frac{\hbar^2}{m_e G (m_{pro})^{5/3} M^{1/3}} \qquad (66)$$

Where m_{pro} and m_e is the proton and electron mass respectively

Just like the Compton wavelength, there must exist another radius for the consistitution of stars that differs from the radius given in (66) above. For example, in the same way the Planck mass is deduced (i.e by equating the Schwarzschild radius to the Compton wavelength) is the same way in which we are to prove the existence of the mass limit of a black hole.

We start from first principles. Let it be known that the derivation of the Chandrasekhar mass limit will follow the equipartition of the gravitational potential energy of a star to its electron degeneracy pressure. Whereby, if the gravitational binding energy is given by,

$$E_g = \frac{2M_{pl}{}^3 m_e c^2 (6.144\pi^3)}{Mm_{pro}{}^2 \quad \mu^2}$$

Where M_{pl}, is the Planck mass and μ is the average molecular weight per electron

And the electron degeneracy energy pressure of the star is given by,

$$E_d = m_e c^2$$

When $E_g = E_d$ then we obtain the mass limit of the white dwarf star as,

$$M = \frac{12.288\pi^3 M_{pl}{}^3}{\mu_e{}^2 \quad m_p{}^2} = 1.4 M_{sun}$$

If then this is true, then the formula for the gravitational binding energy of a star is also true. This therefore implies that the following assumption will also be true.

When the binding gravitational energy of a star is equal to the Newtonian gravitational potential energy $\frac{GM^2}{R}$ we obtain the radius which is the smallest size to which a given mass of a star can be localized as,

$$\frac{GM^2}{R} = \frac{2M_{pl}^3 m_e c^2 (6.144\pi^3)}{M m_{pro}^2 \; \mu^2}$$

$$R = \frac{G m_{pro}^2}{2 m_e c^2} \left(M/M_{pl}\right)^3 \frac{\mu^2}{6.144\pi^3} \qquad (67)$$

This can be rewritten in the form,

$$R = R_k \left(M/M_{pl}\right)^3$$

Where $R_k = 2.384 \times 10^{-53}$ m which is smaller than the Planck length of 1.62×10^{-35} m

Therefore equating Equation (66) to Equation (67) we deduce the mass limit of a black hole as,

$$M = \left(\frac{293.534\pi^{11} M_{pl}^{21}}{\mu^6 M_{pro}^{11}} \right)^{1/10}$$

$$= 9.54 \times 10^{13} \text{Kg}$$

The value is in excellent agreement with other theoretical and experimental observations

The radius of this black hole from Equation (67) is thus 2.527×10^{14} m larger than the radius of the sun of 7×10^8 m.

In conclusion therefore the end stage in the evolution of a star can only be that of the black hole with a mass 9.54×10^{13} Kg and size of 2.53×10^{14} m in contrast with the Chandrasekhar observations.

Note that the radius given by Equation (67), $R = R_k \left(M/M_{pl} \right)^3$ above is similar to the Equation for the size of the Planck star that was given by Rovelli and Vidotto, $r = l_p \left(\frac{M}{M_{pl}} \right)^n$ where l_p is the Planck length and n is the positive number. This is a clear indication that there is a length that is smaller than the Planck length.

On the Quantum Electrodynamics and Quantum Gravity Magnetic Field Limits.

The scale in quantum electrodynamics (QED), above which the electromagnetic field is expected to become non linear, also called the Schwinger limit, was first derived by Fritz Sauter in 1931. However In this section we develop a mechanism (which differs from Fritz's approach) through which the Schwinger limit is deduced using a dimensionless number, which gives the critical magnetic field in quantum electrodynamics when its value is equal to the electromagnetic coupling constant and in the same way gives the critical magnetic field in Quantum gravity when its value is equal to the gravitational coupling constant.

According to D.A. Leahy, the application of quantum electrodynamics in strong magnetic fields only fairly recently has been a subject of interest. The motivation for this study was the discovery of Neutron stars with very high magnetic fields of orders 10^{12}-10^{13}G.

With the discovery of magnetars, quantum electrodynamics calculations which are valid for very high fields became of great interest. The critical value of the magnetic field is defined as $B = \frac{m^2 c^2}{\hbar e} = 4.414 \times 10^{13}$ G. However, there is a value of the magnetic field that is bigger and stronger than the critical magnetic field strength in Quantum electrodynamics and this magnetic field is of orders of magnitude 10^{52} G.

Such a big value has not been deduced in any existing scientific literature and that is the reason why I take pleasure in deriving it here and hence call it the "quantum gravity threshold".

From equation (37) $\frac{F_B c}{A} = \frac{F_e^{\,2}}{2nh}$, if we let the magnetic force to be equal in magnitude and strength to the electric force, $F_B = F_B = Bev$, but since intensity is power per unit area, where the power of the magnetic field is $P = Bevc$, then

$$\text{Force (F)} = \frac{2hc}{A} = \frac{Bev}{n}$$

If n was the fine structure constant ($ke^2/\hbar c$, $k = 1/4\pi\varepsilon_o$), the speed of light in vacuum being $c = \lambda f = \lambda\omega/2\pi$ and the velocity of a particle in the magnetic field is $v = \omega r$ where ω is the angular frequency for circular motion we have

$$\frac{F_c}{F} = \frac{em}{2B\lambda\hbar\varepsilon_o} \qquad 68$$

Where $F_c = m\omega^2 r$ is the centripetal force

We have thus derived a general formula for the coupling of forces. Then the Schwinger limit in quantum electrodynamics for the critical magnetic field can be deduced from the above expression when we set the ratio of the forces to be equal to the electromagnetic coupling or fine structure constant as, $\frac{F_c}{F} = \frac{ke^2}{\hbar c}$

$$B_{QED} = \frac{2\pi mc}{\lambda e}$$

For a particle with deBrogile wavelength $\frac{2\pi\hbar}{mc} = \lambda$, the quantum electrodynamics threshold is given by,

$$B_{QED} = \frac{m^2 c^2}{\hbar e} = 4.3697 \times 10^{13} G \qquad 69$$

However, for $\frac{F_c}{F} = \frac{Gm^2}{\hbar c}$, the gravitational coupling constant, and $\frac{2\pi\hbar}{mc} = \lambda$, the deBrogile wavelength, the quantum gravity threshold is given by a value,

$$B_{QG} = \frac{ec^2}{4\pi G\hbar\varepsilon_0} = 1.8423 \times 10^{52} G \qquad 70$$

We have thus deduced the constant magnetic field carried by an electron in the combined quantum electromagnetic and gravitational fields. The fact that the formula has the fundamental constant of electricity (ε_o), relativistic quantum mechanics (c, $\hbar$) and Gravity (G), is an indication that this is the quantum gravity limit or a scale at which the electromagnetic field is expected to become non linear.

In conclusion, the dimensionless coupling constant is therefore the ratio of the universal constant magnetic field (B_0) on the particle of mass m to the Schwinger magnetic induction limit- the strong magnetic field B external to virtual electron-positron pair enclosed in a quantum vacuum. In general, the coupling constant can be written as,

$$\alpha = \frac{em}{2B\lambda\hbar\varepsilon_o}$$

For $\lambda = \frac{2\pi\hbar}{mc}$, the coupling constant reads as, $\alpha = \frac{em^2c}{4\pi B\lambda\hbar^2\varepsilon_o}$

Implying, $\alpha = \frac{B_o}{B}$

Where $B_o = \left(\dfrac{m}{\hbar}\right)^2 \dfrac{ec}{4\pi\varepsilon_o} = 3.22 \times 10^{11}\,G$,

This value of the magnetic field is smaller than the Schwinger limit and therefore has implications for the theory of QED. Once such a value is discovered in experiments it will lead to a complete observation of the electron-positron pairs. Finally the fine structure coupling constant and the gravitational coupling will manifest in situations where the magnetic field B is either a Schwinger limit given by equations 69 and 70 above. This implies that, the value of the magnetic field B_o derived above is the same everywhere (i.e both in applications of electromagnetism and gravity)

Emergent Gravity

Gravity, Inertia and Electromagnetism as a result of Quantum vacuum fluctuations

Starting from first principles and general assumptions we present a heuristic argument that shows that Newton's law of gravitation and Coloumb's law of electricity naturally arise in a theory in which space emerges through a zero- point fluctuation of the quantum vacuum. Gravity is identified with a casimir force caused by quantum vacuum fluctuations due to the presence of material bodies in it or the distortion of the vacuum through its interaction with mass. A relativistic generalization of the presented arguments directly leads to the Einstein equations. When space is emergent even Newton's law of inertia needs to be explained. The equivalence principle suggests that it is actually the law of inertia whose origin is casimir.

The real origin of gravity is one of the most important, complex and substantially yet unsolved questions in Physics. The replacement of the Newtonian model of gravity with the Einstein's one given by General Relativity (GR) has only shifted the question without solving it. Within GR, gravity has two possible interpretations: a field one and a geometric one. According to the latter, that has become the prevalent one, gravity is due to the curvature of the space – time "tissue", represented as a "rubber sheet", due to the presence of a mass. Nevertheless, this is a purely mathematical description telling nothing about the physical mechanism starting the motion. In fact, even supposing the existence, in the neighbouring of a source mass, of a curved four – dimensional manifold it doesn't explain why a second particle at rest should move towards the source mass.

As such, it invites attempts at derivation from a more fundamental set of underlying assumptions, and six such attempts are outlined in the standard reference book Gravitation, by Misner, Thorne, and Wheeler (MTW). ' Of the six approaches presented in MTW, perhaps the most far-reaching in its implications for an underlying model is one due to Sakharov; namely, *that gravitation is not a fundamental interaction at all, but rather an induced effect brought about by changes in the quantum fluctuation energy of the vacuum when matter is present.* ' In this view the attractive gravitational force is more akin to the induced van der Waals and Casimir forces, than to the fundamental Coulomb force. Although speculative when first introduced by Sakharov in 1967, this hypothesis has led to a rich and ongoing literature on quantum-fluctuation-induced gravity that continues to be of interest.

In this approach the presence of matter in the vacuum is taken to constitute a kind of set of boundaries as in a generalized Casimir effect, and the question of how quantum fluctuations of the vacuum under these circumstances can lead to an action and metric that reproduce Einstein gravity has been addressed from several viewpoints.

Therefore in this chapter we want to show that gravitation might be not a fundamental interaction but a byproduct of the electromagnetic interaction, precisely an electromagnetic phenomena induced by the presence of matter in the quantum vacuum (the quantum field that is present even in empty space). Which means that, matter is not just there but is in the quantum vacuum, and therefore interacts with it, causing some kind of quantum fluctuation energy, that fluctuation is gravitation. In simple terms, a body immersed in quantum fields will interact with them causing gravity to manifest.

Emergence of the laws of Newton

Haisch, Rueda, and others have made the claim that the origin of inertial reaction forces can be explained as the interaction of electrically charged elementary particles with the vacuum electromagnetic zero-point field expected on the basis of quantum field theory.

Gravity is treated as a residuum force in the manner of casimir or vander waals forces. Expressed in the most rudimentary way this can be viewed as follows. The zero point field causes a given charged particle to oscillate. Such oscillations give rise to secondary electromagnetic fields. An adjacent charged particle will thus experience both the zero point field driving forces causing it to oscillate, and in addition forces due to the secondary fields produced by the zero point field driven oscillations of the first particle. Similarly, the zero point field driven oscillations of the second particle will cause their own secondary fields acting back upon the first particle. The net effect is an attractive force between the particles.

Force and Inertia

For the interaction between two particles, each mass experiences a background zero point field and a zero point driven dipole field of the other mass.

Two masses A and D (taken here to be equal for ease of discussion) with D located a distance R from A, along the positive z axis of a coordinate system centered at A. The zero point field will cause a charged particle A to oscillate. The oscillations will then give rise to a secondary electromagnetic field, which will cause particle D to oscillate. In the same way, the zero point field driven oscillations of particle D will cause their own secondary fields acting back upon particle A. the net effect will be an attractive force between particles A and D that will cause one to move towards the other with a small acceleration a_0 in the weak field limit.

Analogous to the Compton Effect, the wavelength of the electromagnetic waves emitted or scattered as a result of particle A interacting with the quantum vacuum will be given as,

$$\lambda_1 = \frac{2\pi R^2}{\lambda_c} \left(\frac{B_0}{B}\right) \alpha$$

Where,

B_0- is the strong magnetic field (greater than or equal to the critical value, which can create electron-positron pairs from the quantum vacuum). The Schwinger mechanism has two cornerstones, the first one is the existence of quantum vacuum and the second one the existence of an external electric field (which attempts to separate electrons and positrons). There are no particles in the vacuum (in that sense the vacuum is empty); but the vacuum is plenty of short-living virtual particle-antiparticle pairs which in permanence appear and disappear (allowed by time- energy uncertainty relation). A "virtual" pair can be converted into a real electron-positron pair only in the presence of a strong external field, which can spatially separate electrons and positrons, by pushing them in opposite directions, as it does an electric field. Therefore the zero point field or quantum vacuum exists but with an external magnetic field stronger than the critical value such that when a particle A is immersed in this zero point field or quantum vacuum, it will interact with the quantum vacuum causing quantum vacuum fluctuations which will trigger the external magnetic field causing oscillation of particle A giving rise to secondary electromagnetic fields.

B- is the value of the magnetic field that exists between particle A and D. This value depends only on the masses of the particles. In other words it is the magnetic field that depends on the matter constituents of the particles in question irrespective of the distance.

$$B = \frac{m^2 ec}{4\pi\varepsilon_o \hbar^2}$$

λ_c- is the reduced Compton wavelength $\frac{\hbar}{mc}$ and α is a dimensionless coupling constant.

It must be noted that, in the weak field limit, the resistance which defines the inertia of a particle is, ultimately, electromagnetic resistance caused by the zero point field on the particle, and it is this resistance which produces gravitational waves with a wavelength of due to a state of motion of a particle,

$$\lambda_2 = \frac{2\pi c^2}{a_o} \alpha$$

From the above given assumption, it is proposed that a body's inertia is due, to the distribution of matter in the universe, and, more precisely, to the electromagnetic interaction that arises from quantum fluctuations of the zero point field in accelerated frames. Basically, a particle's inertia is a function of the particle's interaction with zero point field. Inertia is resistance to acceleration and this reistance causes a form of the gravitational wave simply because resistance becomes a force. This implies that, the resistance which defines the inertia of a particle is, ultimately, electromagnetic resistance caused by the zero point field on the particle.

It must therefore be true that under a condition where, $\lambda_1 = \lambda_2$, we recover Newton's law of inertia (F=ma) as,

$$F = ma_o = \frac{\hbar c}{R^2}\left(\frac{B}{B_o}\right) \qquad (71)$$

Therefore matter continuosly interacts with the zero point field (as Casimir effect), and this interaction yields a force (the resistance to motion) whenever acceleration takes place. Inertia is due to the distortion of the zero point fluctuations in an accelerated reference frame. Technically, inertia is due to the high frequencies of the distortion of the zero point spectrum.

Newton's law of gravity

For the interaction between two masses, each mass experiences a background zero point field and a zero point field driven dipole field of the other mass. The procedure followed here is precisely that developed by Boyer for the derivation of the retarded van der waals forces at all distances between a pair of polarizable particles. Therefore we need only outline the procedure as it applies to our case.

Two masses A and D (taken here to be equal for ease of discussion) with D located a distance R from A, along the positive z axis of a coordinate system centered at A. The modified casimir force between the pair of particles A and D is given by Eqn71,

$$F = ma_o = \frac{\hbar c}{R^2}\left(\frac{B}{B_o}\right)$$

Where B_o is the external (or background) magnetic field stronger than the critical value and B is the dipole magnetic field at the position of particle A due to the motion of particle D and so forth. But since, $B_o = \frac{ec^2}{4\pi\varepsilon_o G\hbar}$, and $B = \frac{m^2 ec}{4\pi\varepsilon_o \hbar^2}$, on substitution into Eqn71 we obtain a familiar law,

$$F = \frac{Gm^2}{R^2}$$

We have recovered Newton's law of gravitation, practically from first principles!

These equations do not just come out by accident. It had to work, partly for dimensional reasons. In a sense we have reversed these arguments. But the logic is clearly different, and sheds new light on the origin of gravity: it is a casimir force! That is the main statement, which is new and has not been made before. If true, this should have profound consequences.

It is hereby proposed that, gravity is not a separately existing fundamental force, but rather a residuum force derived from zero-point fluctuations of other fields in the manner of the Casimir and van der Waals forces. Particularizing this hypothesis to the zero point fluctuation of the vacuum electromagnetic field, we identify the gravitational force as the casimir force associated with the long-range radiation fields (as opposed to the usual shorter-range induction fields) generated by the particle motion response to the zero point fluctuation of the electromagnetic field.

It is therefore seen that a well-defined, precise quantitative argument can be made that gravity is a form of long-range casimir force associated with particle response to the zero-point fluctuations of the electromagnetic field. As such, the gravitational interaction takes its place alongside the short-range van der Waals forces and the Casimir force as related phenomena which emerge from the underlying dynamics of the interaction of particles with the zero-point auctuations of the vacuum electromagnetic field.

Emergence of electromagnetism

Electromagnetism or the coloumb force emerges in a similar fashion as the gravitational force. The origin of the electric force here assumes a critical magnetic field (Schwinger effect or limit) taken here to represent the external magnetic field,

$$B_o = \frac{m^2 c^2}{\hbar e}$$

But since, $= \frac{m^2 ec}{4\pi \varepsilon_o \hbar^2}$. On substitution into Eqn71 we obtain a familiar law,

$$F = \frac{e^2}{4\pi \varepsilon_o R^2}$$

We have recovered Coulomb's law of electricity, practically from first principles!

These equations do not just come out by accident. It had to work, partly for dimensional reasons. In a sense we have reversed these arguments. But the logic is clearly different, and sheds new light on the origin of electricity and gravity: it is a casimir force

Therefore the gravitational field is the set of all electromagnetic fields generated by all particles as they interact with the zero point fields. Gravity and electricity results from a distortion of the quantum vacuum through its interaction with a mass.

A New Alternative to Entropic Gravity

The first attempt into the derivation of the gravitational force and Newton's laws of inertia was given in part by Erik Verlinde (2011) in which he stated that gravity is an entropic force. Simple as it was, his ideas were on a large scale rejected by mainstream physicists. The rejection of verlinde ideas where not backed up by another approach as it has been with other theories save for Sabine Hossenfelder approaches and critics. I think Erik was not surprised by these attacks because this is what physicists do especially when ones analysis or derivation doesn't involve the use of rigorous mathematical models-the one which were used by Einsten and others.

Anyway, what could be another approach towards the derivation of the gravitational, electricity and the law of inertia different from Verlindes idea of the emergent of gravity as an entropic force?

In this chapter we derive Newton's laws of inertia, gravitation and also the electromagnetic force law from first principles without assuming dark matter and the MOND theories. To differ from Verlindes approach we shall use the notions of Quantum vacuum and the Schwinger effect or limit in QED.

In order to understand the physical significance of the derivation to be given herein, we must remember the Schwinger mechanism (Schwinger, 1951) in Quantum Electrodynamics: a strong electric field, greater than a critical value, can create electron-positron pairs from the quantum vacuum.

The Schwinger mechanism has two cornerstones, the first one is the existence of quantum vacuum and the second one the existence of an external electric field (which attempts to separate electrons and positrons). There are no particles in the vacuum (in that sense the vacuum is empty); but the vacuum is plenty of short-living virtual particle-antiparticle pairs which in permanence appear and disappear (allowed by time-energy uncertainty relation). In simple words, the quantum vacuum is a kingdom of the virtual particle-antiparticle pairs; a kingdom with apparently perfect symmetry between virtual matter and virtual antimatter.

A "virtual" pair can be converted into a real electron-positron pair only in the presence of a strong external field, which can spatially separate electrons and positrons, by pushing them in opposite directions, as it does an electric field. Thus, "virtual" pairs are spatially separated and converted into real pairs by the expenditure of the external field energy. For this to become possible, the potential energy has to vary by an amount in the range of about one Compton wavelength, which leads to the conclusion that a significant pair creation occurs only in a very strong external field E, greater than the critical value.

Therefore the external force which attempts to separate particles and antiparticles converting a virtual pair into a real one may be simplified as,

$$F\Phi = 4\pi E_o \hbar \quad (72)$$

Where F is the external force, Φ is the magnetic flux impeding a sphere of radius R and area $A = 4\pi R^2$ and E_o is the electric field on an electron of mass m

Derivation of force and inertia

From chapter 6, the magnetic field is related to the electric field by

$$E_o = B_o c = \frac{Mme\, c^2}{4\pi \varepsilon_o \hbar^2} \quad (73)$$

Remember the above given value is constant for a particle with mass m. There is also an assumption for M=m. Note also that $B_o \neq B$

A new assumption and probably the most surprising one is that, the magnetic flux can be related to energy W and acceleration a by

$$\Phi = \frac{We}{a\hbar\varepsilon_o} \quad (74)$$

Thus, "virtual" pairs are spatially separated and converted into real pairs by the expenditure of the external field energy. For this to become possible, the potential energy has to vary by an amount in the range of about one Compton wavelength, which leads to the conclusion that a significant pair creation occurs only in a very strong external field E, greater than the critical value E_o.

But, $W = Mc^2$, it is therefore evident why equation (74) for the magnetic flux was chosen to be of the given form. It is picked precisely in such a way that one recovers the second law of Newton

$$F = ma$$

As easily verified by combining (72) together with (73) and (74)

Therefore, a similarity or resemblance between acceleration, thermodynamics and electromagnetism comes alive in the statement below;

As there is a formula for the temperature T that is required to cause an acceleration equal to a, $T = \frac{\hbar a}{2\pi kc}$ so there must also be a temperature required to cause an acceleration for an electron in the quantum vacuum at a constant electric and magnetic flux, $T = \frac{\hbar a \varepsilon_o \Phi}{ke}$

Derivation of Newton's law of Gravity

Suppose our universe is a sphere of area $A = 4\pi R^2$ with a sea of virtual particles in a quantum vacuum. It is theorized as before that to separate the virtual particle and antiparticles in a vacuum into real particles one will require a strong external electric or magnetic field B. For the gravitational field, this external magnetic field was calculated in chapter 6Eqn70 to be,

$$B = \frac{ec^2}{4\pi G \hbar \varepsilon_o}$$

Where we introduced a new constant G. Eventually this constant is going to be identified with Newton's constant, of course. But since we have not assumed anything yet about the existence of a gravitational force, one can simply regard this equation as the definition of G.

Then, the magnetic flux will be given as,

$$\Phi = BA = \frac{ec^2 R^2}{G \hbar \varepsilon_o} \quad (75)$$

Substituting Equation (73) and (75) into (72) one obtains the familiar law

$$F = G\frac{Mm}{R^2}$$

We have recovered Newton's law of gravitation, practically form first principles. Following the above derivation carefully, it implies that gravity is a quantum force resulting from the quantum fluctuations of the vacuum due to an existence of an external strong electric or magnetic field separating particles from antiparticles or matter from anti matter.

Derivation of the law of electromagnetism

Following the same steps as in the previous derivation for the gravitational force but assuming a different external magnetic field (Schwinger limit as given in chapter6Eqn69),

$$B = \frac{Mmc^2}{\hbar e}$$

Then, the magnetic flux will be given as,

$$\Phi = BA = \frac{4\pi Mmc^2 R^2}{\hbar e} \quad (76)$$

Substituting Equation (76) and (73) into (72) one obtains the familiar law

$$F = \frac{e^2}{4\pi\varepsilon_o R^2}$$

We have recovered Coulomb's law of electromagnetism, practically form first principles.

Therefore in principle, every external force which attempts to separate particles and antiparticles, may convert a virtual pair into a real one. If it is always an attractive force, as commonly believed today, gravity can't separate particles and antiparticles. Hence, the conjectured gravitational repulsion between matter and antimatter is a necessary condition for separation of particles and antiparticles by a gravitational field and consequently for the creation of particle-antiparticle pairs from the quantum vacuum. But while an electric field can separate only charged particles, gravitation as a universal interaction might create particle-antiparticle pairs of both charged and neutral particles. Thus, the hypothesis of antigravity opens possibility for a gravitational version of the Schwinger mechanism.

In conclusion, gravity is not an entropic force but rather a quantum force stemming from the quantum fluctuation of particles in the vacuum. That is the main statement, which is new and has not been made before. If true, this should have profound consequences.

A new Approach to the Modification of Newtonian Dynamics (MOND)

Hypothesis

From equation (74) we showed that the gravitational acceleration of a particle is related to the kinetic energy W and magnetic flux Φ by the following equation,

$$a_g = \frac{Wq}{\Phi \hbar \varepsilon_o}$$

q-charge on a particle

The above equation implies that, keeping the energy a constant, the acceleration of a particle increases with a fall in the magnetic flux and vice versa is true. Thus the acceleration of a particle is to a great extent affected by the magnetic field.

From the above given relationship we deduce a modification of the Newtonian dynamics in a limit of small accelerations similar to the Milgrom hypothesis (Astrophys. J. 270, 365-1983) but with an electric force as a cause of these small accelerations and we deduce the value of this acceleration in relation to the movement of electrons in the hydrogen atom. Using the same analysis we also deduce the value of the total mass of the universe.

If we let the energy or the kinetic energy, the work done to move the particle around a body of mass M (a galaxy) in the magnetic field created by M, in a process of magnetic induction be given by, $W = mv^2$, where v is the velocity of the particle and m is the mass of the particle. Also if we let the magnetic flux at any point in space at a distance R from M be $\Phi = 4\pi R^2 B$, then the acceleration is simplified in this way,

$$a_g = \frac{mv^2 q}{4\pi R^2 B \hbar \varepsilon_o}$$

This can then be re-written to resemble the Milogram formula in this way,

We know the magnetic field B is related to the velocity and electric field E by, Bv=E, then the acceleration of a particle is,

$$a_g = m \left(\frac{qa}{4\pi \varepsilon_o E \hbar v} \right) a$$

Where, $a = \frac{v^2}{R}$

This then reduces to,

$$a_g = \left(\frac{q^2 a}{4\pi\varepsilon_o hv \left(\frac{Eq}{m}\right)}\right) a$$

This then gives,

$$a_g = \left(\frac{a}{\frac{1}{\alpha_e}\left(\frac{Eq}{m}\right)}\right) a$$

From the above given given equation, $\alpha_e = \frac{q^2}{4\pi\varepsilon_o hv}$ is the fine structure constant. Eq is the electric force on the particle of charge q. Then the acceleration of a particle of mass m in the electric field will be given by, $a_1 = \frac{Eq}{m}$.

Now relating this to the Milogram Hypothesis of $a_g = \left(\frac{a}{a_o}\right) a$, where $a_o \sim 1.2 \times 10^{-10} m/s^2$, we have

$$\left(\frac{a}{\frac{1}{\alpha_e} a_1}\right) a = \left(\frac{a}{a_o}\right) a$$

186

Therefore in a limit where v=c (the speed of light) and q=e (the charge on an electron), we have

$$\alpha_e = \frac{e^2}{4\pi\varepsilon_o \hbar c} = 1/137$$

This implies that, the acceleration of a particle in an electric field of a body of mass M will be given as

$$a_1 = \alpha_e a_o = 8.759 \times 10^{-13} m/s^2 \quad (77)$$

The above given value implies that, in the presence of the electric and magnetic field the Milogram acceleration constant must be corrected to fit the data very well.

Then the modified Newtonian law of gravity is written as,

$$F = \frac{GMm}{\left(\frac{\alpha_e a}{a_1}\right) R^2}$$

In a limit $a_1 = a$ we don't recover the Newtonian law except for $\alpha_e = 1$, which proves to be difficult. This therefore requires us to rethink gravity with a new gravitational constant that could read as $G_B = \dfrac{G}{\alpha_e} = 9.138 \times 10^{-09} Nm^2/kg^2$

Calculation Of The Total Mass Of The Galaxy From Equation (d)

Remember previously in chapter 12 we deduced the ratio of the accelerations to be equal to the finestructure constant as,

$$\frac{g_b}{g_a} = \alpha_e \qquad (78)$$

Where,

$$g_a = \frac{8\pi\varepsilon_o \hbar c^5}{e^2 GM}$$

And,

$$g_b = \frac{c^4}{GM}$$

Relating these accelerations to the ones given in equation d we should be able to deduce the mass of the galaxy as,

$$g_a = \frac{8\pi\varepsilon_o \hbar c^5}{e^2 GM} = a_o = 1.2 \times 10^{-10} m/s^2$$

From which the total galaxy mass is given as,

$$e^2 M = 7.122 \times 10^{18} kg C^2$$

For $e = 1.602 \times 10^{-19} C$, the total mass is given as,

$$M = 5.55 \times 10^{56} kg$$

This is almost the total mass of the Universe.

Note: since the total mass of ordinary matter in the universe is known to be $M = 1.5 \times 10^{53} kg$, then using the above given equation, the charge required to give this value is $e = 6.891 \times 10^{-18} C$

But with,

$$g_b = \frac{c^4}{GM} = \alpha_e a_o = 8.759 \times 10^{-13} m/s^2$$

We have, $M = 1.387 \times 10^{56} kg$

Because this is not the same mass obtained with the use of g_a, we then rewrite the equation of g_b to give the same value of mass as,

$$g_b = \frac{nc^4}{GM}$$

Where, n is the number taking on values from, n=2,4,16,…………

In conclusion, we have not only modified the Newtonian dynamics as a requirement to account for darkmatter but we have just deduced the total mass of the universe. This implies that we are near or probably we have just completed the theory of quantum gravity.

Reinventing Gravity "The Fifth Force"

__The Big problem__: Is dark matter responsible for differences in observed and theoretical speed of stars revolving around the centre of galaxies, or is it something else?

Why would we want to modify Einstein's outstanding intellectual achievement?

a) Newtonian and Einstein gravity cannot describe the motion of the outermost stars and gas in galaxies correctly.

b) If dark matter is not detected and does not exist, then Einstein's and Newton's gravity theories must be modified.

Since the 1970s and early 1980s, a growing amount of observational data has been accumulating that shows that Newtonian and Einstein gravity cannot describe the motion of the outermost stars and gas in galaxies correctly, if only their visible mass is accounted for in the gravitational field equations.

To save Einstein's and Newton's theories, many physicists and astronomers have postulated that there must exist a large amount of "dark matter" in galaxies and also clusters of galaxies that could strengthen the pull of gravity and lead to an agreement of the theories with the data. This invisible and undetected matter removes any need to modify Newton's and Einstein's gravitational theories. Invoking dark matter is a less radical, less scary alternative for most physicists than inventing a new theory of gravity.

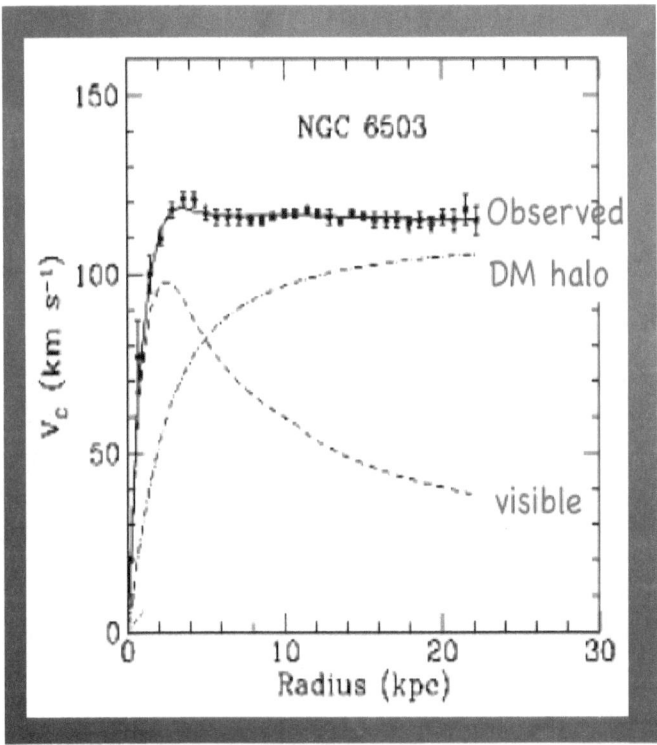

Fig. Galaxy data that show that Newtonian and Einstein gravity do not fit the observed speed of stars in orbits inside a galaxy such as NGC 6503

If dark matter is not detected and does not exist, then Einstein's and Newton's gravity theories must be modified. Can this be done successfully? Yes! My modified gravity (MOG) can explain the astrophysical, astronomical and cosmological data without dark matter.

The Modified Force Law

Let the force of gravity be simplified as,

$$F = p\sqrt{\frac{2\pi\alpha g}{\lambda}} \qquad (79)$$

Where p=mc, is the relativistic momentum of a particle of mass m. α is the coupling constant which determines the strength of a force in any interaction at the range determined by the Compton wavelength λ. $g = \frac{GM}{R^2}$ is the acceleration due to gravity for a point particle at a distance R from a star of mass M.

The above given law was used in chapter 10 and chapter 11 and it was expressed as, $F = \frac{me}{R}\sqrt{\frac{GM\omega}{4\pi\hbar\varepsilon_0}}$

Where $\omega=2\pi f$ is the angular frequency of the graviton-photon oscillations and e is the charge on an electron. In a limit of $\omega = \frac{GM}{R^2}\left(\frac{4\pi\hbar\varepsilon_0}{e^2}\right) = \frac{g_N}{c\alpha_e}$, *where g_N is the usual Newtonian acceleration due to gravity and α_e is the fine structure constant, the above new force law reduces to the Newtonian law of universal gravitation.*

It must be noted that the above force law Eqn1 reduces to the Newtonian law of gravity $\frac{GMm}{R^2}$, only when the following two conditions are met:-

Condition1. The wavelength is equal to the circumference of the circle swept out by the orbiting mass m around M, that is $\lambda = 2\pi R$.

Condition 2. When the coupling constant is half the Newtonian deflection angle θ for a light ray under the influence of a high gravitational field at the sun's surface (α =θ/2=Rs/2R, where Rs is the Schwarzichilds radius)

When the two conditions given above are met, then Eqn79 will become the Newton's law of gravity.

The Tully-Fisher Relation

One of the best fit predictions of MOND is a single universal Tully-Fisher relation.

" The relation between asymptotic velocity and the mass of the galaxy is an absolute one" (Milgrom 1983). This is given by, $V^4 = a_o GM$, where $a_o = 1.2 \times 10^{-10} ms^{-2}$. In this chapter an equation similar to the Tully-Fisher relation is deduced from (79) as given below,

For circular orbits about a mass M, we have the centripetal force equal to Eqn1 as,

$$\frac{mV^2}{R} = p\sqrt{\frac{2\pi\alpha g}{\lambda}} = \frac{mc}{R}\sqrt{\frac{2\pi\alpha GM}{\lambda}}$$

This gives an asymptotically rotation velocity independent of R:

$$V^4 = \left(\frac{2\pi\alpha c^2}{\lambda}\right) GM = a_0 GM \qquad (80)$$

It is this behavior that gives rise to *asymptotically flat rotation curves* and the Tully-Fisher relation (Tully & Fisher 1977) without invoking dark matter.

Comparing (80) to the Tully-Fisher relation, we determine the acceleration limit as,

$$a_0 = \frac{2\pi\alpha c^2}{\lambda}$$

From which the coupling constant takes on values of,

$$\alpha = \frac{a_0 \left(\frac{\hbar}{mc}\right)}{2\pi c^2}$$

This result implies that, if the milgrom acceleration was really a constant of $a_o = 1.2 \times 10^{-10} ms^{-2}$ and also the fine structure constant was the coupling constant of 1/137, then it will be true that the wavelength or the range of the interaction will be given by exactly ,$\lambda = 3.44 \times 10^{25} m$ which is within the acceptable size of the galaxies. This could help us connect quantum mechanics with gravity at small accelerations.

We anticipate that MOG will modify how stars collapse and the nature of black holes.We know that a supermassive object with mass ~ 3 X 10⁶ MSUN is at the center of our Galaxy (MILKY WAY). We are not able to determine yet whether the object is a GR black hole with a horizon. Perhaps future telescopes and space missions will be able to get close enough to the supermassive object to tell whether it is a black hole in spacetime or some other kind of object. However, as distant observers, we can never see a black hole event horizon form! The formation of the event horizon occurs in the infinite future, so we cannot actually ever see a black hole event horizon form as a star collapses.

Resolving the Proton Radius Puzzle

Today the proton radius is measured via three methods that is, the spectroscopy, nuclear scattering and muonic hydrogen (2010 experiment) methods.

The spectroscopy method uses the energy levels of electrons orbiting the nucleus. This method produces a proton radius of about 8.768×10^{-16} m, with approximately 1% relative uncertainty.

The nuclear method is similar to Rutherford's scattering experiments that established the existence of the nucleus. Small particles such as electrons can be fired at a proton, and by measuring how the electrons are scattered, the size of the proton can be inferred. Consistent with the spectroscopy method, this produces a proton radius of about 8.775×10^{-16} m.

The muonic hydrogen 2010 method by Pohl et al. is similar to the spectroscopy method. However, the much higher mass of a muon causes it to orbit 207 times closer than an electron to the hydrogen nucleus, where it is consequently much more sensitive to the size of the proton. The resulting radius was recorded as 8.42×10^{-16} m. This newly measured radius is 4% smaller than the prior measurements, which were believed to be accurate within 1%.

The discrepancy between the measured values of the proton radius by the methods given above is what is called the proton radius puzzle and the discrepancy might be due to new physics, or the explanation may be an ordinary physics effect that has been missed. In what follows, I deduce the spectroscopy and muonic radius of the proton and thereafter provide a reason for the discrepancy.

In Newtonian law of motion for a body to orbit around another body, there must be a centripetal acceleration $\frac{v^2}{R}$ to keep the body in orbit. Where v is the velocity and R is the distance from the center. In the Hydrogen atom this is different the acceleration is given by,

$$a = \frac{c^2}{R_t}$$

Where c is the constant speed of light and $R_t = \frac{l_p^2}{d}$

l_p^2 is the planck area and d is the Schwarzschild radius of the proton (the radius of the event horizon of a proton black hole).

Then the acceleration of an electron will follow the inverse square law given by,

$$a = \frac{dc^2}{l_p^2}$$

We know that the total electric potential energy of an atom is given by,

$$V = \frac{e^2}{4\pi\varepsilon_o d}$$

Eliminating d from the potential we have,

$$V = \frac{e^2 c^2}{4\pi\varepsilon_o a l_p^{\,2}}$$

We have therefore created the potential that falls off as the acceleration.

By measuring the energy released when the excited electrons fell back to lower-energy states, the Rydberg constant could be calculated, and from this the proton radius inferred.

Since the energy of the photon $E = \frac{2\pi\hbar c}{\lambda}$ released is equal to the total potential energy of an atom we have the radius of the proton as,

$$r_p = \frac{\lambda}{2\pi} = \frac{l_p^2 a}{c^2 \alpha_e}$$

Where, λ is the wavelength and $\alpha_e = \frac{e^2}{4\pi\varepsilon_0 \hbar c}$ is the electromagnetic coupling constant.

Assuming an acceleration that was earlier given in chapter 14 of,

$$a = \frac{2nc^2}{R_s}$$

Where n is positive coupling number and $R_s = \frac{2Gm_p}{c^2}$ is the proton Schwarzschild radius. Then on subsititution into the proton radius formula we have,

$$r_p = \frac{2nl_p^2}{\alpha_e R_s}$$

From the above formula we deduce that,

(i) When n=0.03038 for electrons orbiting the nucleus, we get a proton radius of about 8.768×10^{-16} m, which is the exact proton radius that was produced by the spectroscopy method.

(ii) When n=0.0292 for electrons orbiting the nucleus, we get a proton radius of about 8.42×10^{-16} m, which is the exact proton radius that was produced by the 2010 experiment by Pohl et al.

We notice that the difference between the two values of n for the methods given above is the anomalous magnetic dipole moment given by,

$$\Delta n = n_1 - n_2$$

$$\Delta n = 0.03038 - 0.0292$$

$$\Delta n = 0.00118$$

Therefore the discrepancy in the radius of the proton by the two methods is given by,

$$\Delta r = \frac{2l_p^2}{\alpha_e R_s} \Delta n$$

$$\Delta r = 0.3406 \times 10^{-16} m$$

In conclusion the discrepancy is due to the anomalous magnetic dipole moment and it will never go away.

The Bekenstein Hawking Area-Entropy Law

The Unification of Quantum Mechanics and General Relativity into a Quantum theory of Gravity is one of the great scientific challenges of this generation. A definitive resolution will require solving one of the major problems of Quantum Gravity and that is, the Bekenstein-Hawking area-entropy law, $S = a\frac{Ac^3k}{\hbar G}$ (81), where A is the surface area of the Schwarzschild black hole, a is the constant of the order of unity, c is a constant speed of light, k the Boltzmann constant, $\hbar$ the reduced Planck constant and G is the Newton's gravitational constant.

Attempts towards this were done in the early 70s by Hawking who proved that a black hole emits thermal radiation with a temperature $T = \frac{\hbar c^3}{8\pi Gk}$ (82). According to Carlo Rovelli (Dec, 2003), Hawking beautiful result raises a number of questions. First, in Hawking's derivation the quantum properties of gravity are neglected. Are these going to affect the result? Second, we understand macroscopical entropy in statistical mechanical terms as an effect of the microscopical degrees of freedom. What are the microscopical degrees of freedom responsible for the entropy? Can we derive (81) from first principles? Because of the appearence of $\hbar$ in (81), it is clear that the answer to these questions has since become standard benchmark against which a quantum theory of gravity can be tested.

This book presents a simple universal explanation of Black hole thermodynamics in a somewhat different form than that given by Loop Quantum Gravity (LQG), String theory and Hawking radiation theory. The major result of the book is the derivation of (81) from first principles using different methods for Schwarzschild and for other black holes, with a well defined calculation where no infinities appear. As far as this book is concerned there is no other theory from which such a calculation can proceed. Hence the book is the only one from which a detailed quantum theory of gravity precedes and where the result of the Bekenstein-Hawking area entropy law can be achieved.

First Method

In this method we reduce the famous Einstein field equation ($G_{\mu\nu} + \Lambda g_{\mu\nu} = \frac{8\pi G}{c^4} T_{\mu\nu}$ where, the expression on the left represents the curvature of space time while the expression on the right represents the matter-energy content of the universe) to,

$$\frac{1}{R^2} = \frac{8\pi G}{c^4} P_{eg} \qquad (83)$$

Where, R is the radius of a body of mass M, $P_{eg} = \sigma_m \frac{f_e f_g}{\hbar c}$ is the Pressure-Energy density relationship with the coupling of mass (the ratio of the atomic mass, m to the Planck mass M_{pl}) and the electric f_e and gravitational force f_g.

The ratio, $\sigma_m = \dfrac{m}{M_{pl}}$ is introduced to correct for particles approaching the Planck length scale $m \to M_{pl}$

What is the total electric potential energy of a black hole? From (83), we could let the potential electric energy be,

$$E_e = f_e r = \frac{\hbar c^5}{8\pi G E_g \sigma_m}$$

We know that at the Schwarzichild radius $R = \dfrac{GM}{c^2}$, the gravitational potential energy will be of order $E_g = mc^2$, giving the electric energy from (81) as,

$$E_e = \frac{\hbar c^3}{8\pi G M \sigma_m}$$

What is the temperature of a Black hole? Since the thermal energy is given by $E_{thermal} = kT$, where k is the Boltzmann constant

By the principal of Equipartition

$$E_{thermal} \sim E_e \Rightarrow T = \frac{\hbar c^3}{8\pi GMk\, \sigma_m} \tag{84}$$

For $\sigma_m = 1$, we get the usual Hawking temperature of

$$T = \frac{\hbar c^3}{8\pi GMk} \tag{85}$$

We know that, entropy is energy divided by temperature. Having derived the temperature, What is the total energy of a Black hole?

Assuming a law which states that the intensity of the emitted radiation increases as the square of the electric force $I = \beta F_e^2$, where the constant $\beta = \frac{1}{4\hbar}$,

But we can also write the intensity in terms of energy as, $I = \frac{E_T}{tA}$, where t is time and A is the surface area of Schwarzschild black hole. The total energy of a Black hole will then be given as,

$$E_T = \frac{F_e^2 tA}{4\hbar}$$

Let the time taken by a Black hole to evaporate be, $= \dfrac{Mc}{F_e}$, F_e is known from (83), since from the Newtonian law of gravity $F_g R^2 = GM^2$ we then have the total energy of a black hole as,

$$E_T = \dfrac{Ac^6}{32\pi \sigma_m G^2 M} \tag{86}$$

And Power is given by $P = F_e c = \dfrac{\hbar c^6}{8\pi \sigma_m G^2 M^2}$

Then the entropy of a Black hole is given by

$$S = \dfrac{E_T}{T}$$

Substituting in (86) and (84) we obtain the Bekenstein-Hawking area entropy law,

$$S = \dfrac{Ac^3 k}{4\hbar G} \tag{87}$$

Where the constant a=1/4

Second Method

The mathematical ideas behind the theory of general relativity and quantum mechanics are complicated and thus difficult to grasp in all perspectives and this could be one of the reasons why it has been and is still difficult to prove a complete quantum theory of gravity for most physicist. It is in the same domain that I also find the derivations behind the Hawking radiations in literature tiresome and at the same time incomplete. This is why I find it relevant to derive the temperature and entropy and other relations of Black hole physics from first principles, which I think will shed light on the future of quantum gravity and a requirement for the re-development of quantum mechanics and general relativity approaches in a much simpler way as I have labored to deduce it here under three approaches.

Analogous to Einstein's field equation in the theory of General Relativity we create a formula where the scalar curvature is related to the classical Newtonian gravitational and electric forces by,

$$\frac{1}{R^2} = S \frac{F_g F_e}{\hbar c} \qquad 88)$$

Where R is the radius of curvature

F_g is the gravitational force

F_g is the electric force

$\hbar$ is the reduced planck constant

c is the constant speed of light

$s = \frac{8\pi G}{c^4}$, G is the Newton's gravitational constant

From the above expression it is seen that, the expression on the right is the pressure and energy density of matter treated quantum relativisticaly in units of $\hbar c$, as, Pressure = energy density = $\frac{F_g F_e}{\hbar c}$.

Temperature

Let the work done by the electric force to move a particle through a distance R be given by,

$$W = F_e R$$

But the electric force can be got from equation 88, hence giving the work done as,

$$W = \frac{\hbar c}{s F_g R}$$

and the Newtonian force can be given by, $F_g = \frac{Gm^2}{R^2}$ hence deducing the work done as,

$$W = \frac{\hbar c R}{s G m^2}$$

For a black hole the size of the schwarzichild's radius $R = \frac{Gm}{c^2}$, we obtain

$$W = \frac{\hbar}{smc}$$

Assuming that the work done is analogous to the thermal kinetic energy of an ideal gas, W=kT, where, k is the Boltzmann's constant and T is the temperature of the particle, then the temperature of the radiated energy or radiation from the black hole will be given as,

$T = \frac{\hbar}{smck}$ or, $T = \frac{\lambda}{sk}$, where $\lambda = \frac{\hbar}{mc}$ is the deBrogile-wavelength of a particle moving at the spaeed of light. Thus the temperature of the radiated energy from the black holes increases with the wavelength of the emitted particle.

Entropy

Let the entropy be, $S = \frac{E}{T}$, and let the total energy radiated E be given by,

$$E = \frac{F_e^2}{4\pi\hbar} tA$$

Where, t is the time taken by a black hole to dissipate and A is the surface area of the black hole event horizon.

But since the electric force is known from equation 88, then

$$E = \frac{\hbar c^2}{4\pi S^2 R^4 F_g{}^2} tA$$

Let time be, $t = \frac{mc}{F_e} = \frac{ms R^2 F_g}{\hbar}$

Then the total energy will be given by

$$E = \frac{mc^2}{4\pi s R^2 F_g} A$$

Since $F_g = \frac{Gm^2}{R^2}$

Then, $E = \frac{1}{4\pi s R_s} A$, where R_s is the schwarzichild's radius $R_s = \frac{Gm}{c^2}$

Then entropy can be deduced since T is known as, $T = \frac{\lambda}{sk}$, then dividing E by T we obtain,

$$S = \frac{Ak}{4\pi R_s \lambda}$$

If a black hole was a particle with mass m, we can easily compute it's radius and it's wavelength, combining the two, one can compute a new surface area as $A_s = 4\pi R_s \lambda$, then the entropy of a black hole will be given by,

$$S = \frac{A}{A_s} k,$$

If we let the thermodynamic probability be W, then the probability for work done in the expansion from A_s to A is, $W = e^{\frac{A}{A_s}}$.

Third Approach

We write a set of formulas from which our derivations will proceed

1) It is well known that the electric field is force per unit charge but here a generalized equation for an electric field created by an electron exhibiting wave properties in the nucleus of an atom in the gravitational field on a quantum scale is given by

$$E = \frac{1}{r}\sqrt{\frac{Gm^3 f}{2\hbar\varepsilon_0}} \qquad 89$$

Then the electric force in this case will be formulated as

$$F_1 = \frac{e}{r}\sqrt{\frac{Gm^3 f}{2\hbar\varepsilon_0}} \qquad 90$$

2) The surface area at a radius r of orbit of an electron of mass m around the nucleus of an atom in a wave like manner is given by

$$\text{surface area}(A) = \frac{\lambda\mu_0 e^2}{m} \qquad 91$$

3) The time taken by the magnetic field B of an electron to pass through a given surface is

$$\text{time}(t) = \frac{\lambda \varepsilon_0 AB}{e} \qquad 92$$

Note: the above expression is the same as Faraday's induction law.

4) The gravitational force acting on all matter in the universe or the modified gravitational force is given as

$$F_2 = \left(\frac{Gm^3}{r^2}\right)\left(\frac{e}{2B\lambda\hbar\varepsilon_0}\right) \qquad 93$$

The above formulas are important in deriving the formula for the temperature, entropy and the time taken by a black hole to evaporate as shown below;

Temperature of a Black Hole

It is known that the kinetic energy KE of molecules in the Boltzmann hypothesis is related to the temperature of the body in question in this case a black hole (in relation to the black body) by $KE = \varphi T$ where φ is Boltzmann's constant. The formula for the kinetic energy can be derived by using a hypothesis that the electromagnetic force – coulombs force is equal to eqn3 as

$$\frac{ke^2}{r^2} = \frac{e}{r}\sqrt{\frac{Gm^3 f}{2\hbar\varepsilon_o}}$$

On squaring both sides of the equation, cancelling like terms and taking into account that the frequency of an electron is $f = \frac{v}{\lambda}$, then the kinetic energy of an electron inside the black hole is given by

$$KE = \frac{\lambda\mu_o e^2}{A}\frac{c^3\hbar}{8\pi Gm^2}$$

Since the surface area is given as from eqaution91 then the kinetic energy of molecules or particles (for an ideal gas) within the black hole will be given by

$$KE = \frac{c^3\hbar}{8\pi Gm} = T\varphi \qquad 94$$

Then from Boltzmann's relationship the temperature of the black hole is formulated as

$$T = \frac{c^3\hbar}{8\pi Gm\,\varphi} \qquad 95$$

The Entropy of the Black Hole

By definition entropy is a measure of disorder. To solve the entropy of black holes we shall consider a very complex argument about the entropy in question. We assume that the modified gravitational force given by equation6 is identical to the modified electric field given by equation90 as, $\left(\frac{Gm^3}{r^2}\right)\left(\frac{e}{2B\lambda\hbar\varepsilon_0}\right) \equiv \frac{e}{r}\sqrt{\frac{Gm^3 f}{2\hbar\varepsilon_0}}$ in otherwise the two forces are equal but opposite. Then squaring both sides of the equation and multiplying through by Gc^5 one obtains a new relation of forces on both sides given as

$$\frac{tc^7}{16\pi G^2 m} = \frac{Ac^6}{32\pi r m G^2}$$

Both the left and right hand side represent a force. From the left hand side t is the expression of time given by $t = \frac{\hbar e^2}{2m^3 c^2 G\varepsilon_0}$. Note: the left hand side force is the pull of matter inside the black hole while the right hand side force is the force acting on particles or matter at the surface of the black hole.

Since the heat is the product of the force on a particle and the distance r from the centre of the black hole, then using the force on the right hand side of the above equation the heat will be given by

$$Q = \frac{Ac^6}{32\pi mG^2}$$

Remember the temperature of the black hole is also known from equation6 and by definition the entropy of the system is the change in heat per unit temperature $\frac{Q}{T}$, then the entropy of the black hole will be given by

$$S = \frac{A\varphi c^3}{4G\hbar} \qquad 96$$

This implies that the entropy of a black hole is proportional to its surface area.

The Time Taken by a Black Hole to Evaporate

Assuming that particles that formed a black hole are moving away or are separating from it after a given time of its existence, if we measure the relative speed of these particles in relation to the energy they carry we obtain a relationship given by

$$\frac{v^2}{c^2} = \frac{8\pi G}{c^2}\left(\frac{W}{8\pi r}\right) \qquad 97$$

Where v is the velocity of these particles as measured relative to the speed of light c and W is the energy carried by the particles as they move away from the centre of the black hole at a distance r.

If we let the force causing the particles to separate from the black hole be given as $\dfrac{Gm^3 e}{2r\lambda B \hbar \varepsilon_o} \dfrac{v}{c}$, then the energy of these particles will be given by

$$W = \dfrac{Gm^3 e}{2r\lambda B \hbar \varepsilon_o} \dfrac{v}{c}$$

Substituting this in equation 95, we obtain a relationship of time as given by the law 3 of equation 92 as

$$t = \dfrac{v^2}{c^2}\left(\dfrac{\pi G^2 m^3}{\hbar c^4}\right)$$

The velocity of the particles in the astronomical lab will be measured as v= 4.193E6 m/s and since the speed of light is a constant then the time taken by a black hole to evaporate is given by

$$t = \frac{5120\pi G^2 m^3}{\hbar c^4}$$

Fourth Approach

Temperature of a black hole

It is here by hypothesized that, the gravitational field will create particles and emit them only if the electromagnetic force of such particles were equal to the force (unknown in literature) $F = \frac{Me}{r}\sqrt{\frac{Gp}{2\hbar\varepsilon_o\lambda}}$. Where p, is the momentum of a particle. under general conditions, the force given will reduce to the Reissner- Nordstrom metric as given here, if the momentum of an electron at a distance r from the singularity point to the event horizon is related to the de Brogile wavelength as $p = \frac{2\pi\hbar}{\lambda}$, and both the distance r and wavelength λ was the product of the speed of light c and the period T as r=cT and $\lambda = cT$, then the force will be given by $F = \frac{Mp}{r\hbar}\sqrt{\frac{Ge^2}{4\pi\varepsilon_o}}$, but since $\frac{p}{2\pi\hbar} = \frac{1}{\lambda}$, then we have, $F = \frac{2\pi M}{T^2}\sqrt{\frac{Ge^2}{4\pi\varepsilon_o c^4}}$, this reduces to $F = \frac{2\pi M}{T^2}r_q$, where $r_q = \sqrt{\frac{Ge^2}{4\pi\varepsilon_o c^4}}$ is the Reissner-Nordstrom radius of a charged black hole.

Having derived the Reissner-Nordstrom metric from our force formula, we now return to our exercise of deriving the temperature of a black hole. We consider a particle with charge e, exhibiting deBrogile wave properties of momentum and wavelength from the centre of mass M of a black hole. We then assume that this particle experiences an electromagnetic force due to the magnetic and electric field created by other particles in its surrounding area. The same particle also experiences a force due to the strong gravitational field emanating from the black hole. Equating the two forces as $\frac{Me}{r}\sqrt{\frac{Gp}{2\hbar\varepsilon_o\lambda}} = \frac{e^2}{4\pi\varepsilon_o r^2}$, from this expression we obtain the momentum of a particle as $p = \frac{\hbar e^2 \lambda}{2\pi A \varepsilon_o GM^2}$. This is the momentum possessed by a particle (emitted by the gravitational field of a black hole) at the surface of the event horizon, where $A = 4\pi r^2$ is the spherical surface area of the horizon.

For relativistic effects, the kinetic energy of a particle will be related to its momentum by K.E=pc and to the Boltzmann's law by K.E=kT, where k is the Boltzmann's constant and T is the absolute temperature. By similarity we can equate the two energies as pc=kT, then from the equation of momentum we can obtain the temperature as,

$$T = \frac{\hbar e^2 \lambda c}{2\pi A \varepsilon_o GM^2 k}.$$

Expressing the permittivity of free space in terms of the permeability of free space $\varepsilon_o = \frac{1}{\mu_o c^2}$, we obtain the Hawking temperature of a black hole as,

$$T = \left(\frac{4e^2 \mu_o \lambda}{AM}\right)\frac{\hbar c^3}{8\pi GMk}$$

In a more general form, in terms of energies it can be expressed as,

$$T = \left(\frac{4e^2\lambda}{A\varepsilon_0 Mc^2}\right)\frac{\hbar c^3}{8\pi GMk} \qquad 98$$

We propose that, $mc^2 \geq \frac{4e^2\lambda}{A\varepsilon_0}$ and if $A = 4\pi R^2$ and $\lambda = \frac{R}{4}$ then, $mc^2 \geq \frac{e^2}{4\pi\varepsilon_0 R^2}$ the electric potential energy.

Entropy of a black hole

In an attempt to prevent the violation of the generalized second law of thermodynamics, Bekenstein proposed a universal upper bound on the ratio entropy to energy for bounded systems (Phys RevD23, 287-1981), which was later rejected by Unruh and Wald in 1982. They proposed a thought experiment in which a box lowered down into a black hole felt an effective buoyancy force which was caused by the acceleration radiation felt by the box near the black hole. They argued further that, this buoyancy force would guarantee a lower bound on the energy gain of the black hole, hence saving the generalized second law without a need for entropy bound.

In this section we give a formula for the buoyancy force which is different from the Unruh and Wald formula which appeared in their 1982 paper.

At a distance r from the center of mass m of a black hole, the buoyancy force is given by,

$$F_B = \frac{rc^6}{8G^2 m} \qquad 99$$

From the above force formula the energy gain by the black hole will be given by,

$$W_B = \frac{Ac^6}{32\pi G^2 m}$$

Where, A is the area of the event horizon. Since entropy is the ratio of energy to temperature, $S_B = W_B/T_B$ and temperature of a black hole is known from equation 11, then the entropy of a black hole is given by,

$$S_B = \frac{Akc^3}{4G\hbar}\left(\frac{A\varepsilon_0 Mc^2}{4e^2\lambda}\right) \qquad 100$$

New Physics: Regularization and Physics beyond the Standard Model

History tells us that if we hit upon some obstacle, even if it looks like a pure formality or just a technical complication, it should be carefully scrutinized. Nature might be telling us something, and we should find out what it is (G. t Hooft, 1997).

In physics, one of the ultimate goals is to unify the fundamental forces of nature. Today physicists have been able to unify three of the four known fundamental forces (the electromagnetic, the strong and the weak nuclear forces in a single quantum field theory-the standard model). The fourth fundamental force, gravity, on the other hand is described by the general theory of relativity. Because the other fundamental interactions are quantized, it therefore seems natural that in a grand unified theory, a theory of all the fundamental forces, gravity is quantized as well into perhaps Quantum gravity.

A theory of quantum gravity is needed to describe things that are very small but also very heavy, like black holes or the early universe. However, the development of a quantum theory of gravity seems difficult on grounds that, in general relativity all physical qualities have definite values, whereas in quantum mechanics they do not as shown in Heisenberg's uncertainty principle.

The problems in General Relativity arise from trying to deal with a universe that is zero in size (infinite densities). But quantum mechanics suggests that there may be no such thing in nature as a point in space-time, implying that space-time is always smeared out, occupying some minimum region. The minimum smeared-out volume of space-time is a profound property in any quantized theory of gravity and such an outcome lies in a widespread expectation that singularities will be resolved in a quantum theory of gravity.

However, Prof Brian Dolan at the Department of Theoretical Physics, NUI Maynooth, is quick to point out that there is not yet any set agreement on what a theory of quantum gravity should look like, or even on the exact problem it is trying to solve."There is no accepted theory of quantum gravity," he says. "There are currently a number of contenders, and by far the most popular is superstring theory. Many physicists find superstring theory compelling due to its internal elegance, but despite decades of intense research it has not produced a single experimentally testable result." He suspects that trying to unite general relativity and quantum mechanics may be the wrong way to go, and that any future breakthrough may come from a completely unexpected direction; perhaps from some young mind with a fresh perspective.

This chapter employes new idea towards the development of a quantum theory of gravity in a bid to solve the following unsolved problems in physics;

(i) Is it true that at every spatial dimension, there exists new physics and that it is the work of Physicists to find out? What is the method or procedure through which new physics can be found? Does this require extra dimensions?

(ii) Does nature have more than four space-time dimensions? If so, what is their size? Are dimensions a fundamental property of the universe or an emergent result of other physical laws? Can we experimentally observe evidence of higher spatial dimensions?

(iii) Can the singularities that plague the General theory of Relativity be resolved in any quantum theory of Gravity?

The Standard Model is inconsistent with that of general relativity, to the point that one or both theories break down under certain conditions (for example within known spacetime singularities like the Big Bang and the centers of black holes beyond the event horizon).

The appearance of singularities in any physical theory is an indication that something is wrong and that there is a need for new physics. Singularities can be avoided in GR and any field theory through the introduction of an efficient regularization procedure as this book directs.

Regularization is a method of modifying observables which have singularities in order to make them finite by the introduction of a suitable parameter called regulator. The regulator, also known as a "cutoff", models our lack of knowledge about physics at unobserved scales (e.g. scales of small size or large energy levels). **It**

compensates for the possibility that "new physics" (beyond the SM) may be discovered at those scales which the present theory is unable to model, while enabling the current theory to give accurate predictions as an "effective theory" within its intended scale of use.

The need for regularization terms in any quantum field theory of quantum gravity is a major motivation for Physics beyond the standard model. Infinities of the non-gravitational forces in QFT can be controlled via renormalization only but additional regularization and hence new physics is required uniquely for gravity. The regularizers model, and work around, the breakdown of QFT at small scales and thus show clearly the need for some other theory to come into play beyond QFT at these scales. A. Zee (Quantum Field Theory in a Nutshell, 2003) considers this to be a benefit of the regularization framework, theories can work well in their intended domains but also contain information about their own limitations and point clearly to where new physics is needed.

Therefore the main objective of this section is to discover new physics at those scales (or extra dimensions) which the General relativity theory and Quantum mechanics is unable to model. The section also sets out to prove that due to quantum gravitational effects, there is a minimum distance beyond which the force of gravity no longer continues to increase (operate) as the distance between the masses become shorter.

General Theory

During the years, strong evidence has appeared that the acceleration of any physical object cannot be arbitrarily large, but it should be superiorly limited. For example in string theory, it was derived that string acceleration must be less than some critical value, determined by the string tension and its mass. From the classical point of view (as Wheeler suggested), if we consider an extended object in **rotating motion**, we have the acceleration $a = v^2/R$ and it follows that a, must be at least limited by c^2/R. However to differ from the classical Newtonian mechanics and Einstein's General relativity theory we introduce a regulator "Cutoff" $\alpha_g{}^n$, where α_g is the gravitational coupling constant, R is the distance between two masses and n is a positive number (**extra dimension** number), then the acceleration must be limited by $a = \frac{c^2}{2R}\alpha_g{}^n$ (101), (Assuming a diameter of 2R).

Thus to avoid the infinity but while retaining the point nature of the particle would be to postulate a small additional dimension **n** over which the particle could 'spread out' rather than over 3D space.

For example, in the Unruh temperature we can only and only deduce both the Hawking temperature and maximal temperature (Sakharov Temperature) under the assumption of the existence of a maximal acceleration given in formula (101) above as,

The Unruh temperature is given as,

$$T = \frac{\hbar a}{2\pi ck}$$

Since the acceleration is known from (101) above, then the temperature will reduce to,

$$T = \frac{\hbar c}{4\pi Rk} \alpha_g^{\ n}$$

For a Schwarzschild Black hole of radius $R = \frac{2GM}{c^2}$, the temperature reduces to

$$T = \frac{\hbar c^3}{8\pi GMk} \alpha_g^{\ n}$$

Since the gravitational coupling constant has a formula $\alpha_g = \frac{GM^2}{\hbar c}$, taking values of n=0,1,2,............,N. Then the Hawking temperature will become a result of n=0 extra spatial dimensions as,

$$T = \frac{\hbar c^3}{8\pi GMk}.$$

Also the maximum temperature (Sakharov temperature) is deduced at n=1/2 as,

$$T = \frac{1}{8\pi k}\left(\frac{c^5 \hbar}{G}\right)^{1/2}$$

Therefore the temperature of a black hole increases as a black hole loses mass in Hawking Black hole evaporations. The analysis given above is a clear indication that the temperature doesn't increase exponentially as it has been known from Hawking's original proposals, there is a maximum temperature, a limit on temperature that screens (resolves) the classical singularity. It is therefore true that the radiation spectrum contains all Standard Model particles, which are emitted on our brane, as well as gravitons, which are also emitted into the extra dimensions. It is expected that most of the initial energy is emitted during this phase in Standard Model particles. Therefore we recommend the applications of a factor α_g^n in situations involving the examination and experimentation of quantum gravitational phenomenon. We shall see in the coming chapters that such a factor when used in loop quantum cosmology it reproduces both the results of loop quantum gravity and string theory.

The idea of including extra dimensions, to achieve the goal of unifying physics, is not a new one. Already the year before Einstein in 1915 introduced his theory of general relativity; Gunnar Nordstrom suggested a unification of gravity and electromagnetism with the introduction of a fifth dimension. These forces were the two only forms of interaction known at that time. But this idea was forgotten for some time with the eruption of the First World War. But in April 1919 Theodor Kaluza introduced independently, in a letter to Einstein, a fifth dimension in an attempt to unify Einstein's theory of gravity and Maxwell's theory of light. Oskar Klein (1926) contributed, in this quest, with his assumption that the extra dimension was compactified. The Kaluza-Klein theory was a fact. This theory includes an extra space dimension that is rolled up into a tiny circle, i.e. compactified. And in this five dimensional theory, there is only one underlying force, gravity. But in the four-dimensional spacetime observed at great distances, it appears to be three kinds of forces, among these a gravitational and an electromagnetic force. This topic was initially a popular topic for research, but lost much of its interest with the introduction of quantum mechanics.

In recent years the topic of extra dimensions has experienced a renewed interest. This renewed interest is also due to the exciting possibility of observing new and spectacular physical phenomena at far lower energy scales than otherwise. Even at energies available in the not so distant future, these phenomena could appear. Among these is the creation of higher dimensional semi-classical microscopic black holes. The possibility of observing these objects, is viewed as an opportunity to perhaps discover new intriguing physics.

Therefore from (i) using Einstein's equivalence principle we get the minimum distance beyond which the force of gravity no longer continues to increase as;

$$R = \frac{R_s}{\alpha_g{}^n} \qquad (102).$$

Where $R_s = \frac{2GM}{c^2}$ is the Schwarzschild radius. We therefore conclude that;

(i) At n=0 extra spatial dimension, we have a physical theory of General relativity at a length scale of $R = R_s = \frac{2GM}{c^2}$ - the Schwarzschild radius.

(ii) At n=1/2 extra dimension, we have the quantum theory of gravity (New physics) at the Planck length scale $l_p = \sqrt{\frac{\hbar G}{c^3}}$

(iii) At n=1 extra dimension, we have the theory of Quantum mechanics at the Compton wavelength scale of $\lambda = \frac{\hbar}{mc}$.

(iv) Lastly at n=2 we have new physics at a length scale $R = \frac{\hbar^2}{GM m^2}$ and the journey continues.

According to the Standard Model of particle physics, the world is governed by four fundamental forces: gravity, electromagnetism, and the weak and strong nuclear forces. Although things act a bit "spooky" down on the quantum level, science has managed to generally describe all of these forces at both the macro and quantum scales – except gravity.

Gravity is the weakest of the fundamental forces, and it's been suggested that this is because some gravitons (the hypothetical particles) that carry the gravitational force tend to escape into extra dimensions. We're simply too big to travel through or even notice these other dimensions.

So, to study whether these extra dimensions are lurking in extremely tiny spaces, the researchers from Osaka, Kyushu and Nagoya Universities set out to test gravity on the sub nanometer scale. To do so, they used the world's highest intensity neutron beam, which is housed at the Japan Proton Accelerator Research Complex (J-PARC).

The team found that the results matched predictions based on the known laws of physics, which indicates that Newton's law still applies as expected down to a scale of less than 0.1 nanometers. No unexplained force ie, another dimension is acting on these particles at this scale.

That doesn't mean those extra dimensions aren't there, just that they may be hiding at even smaller scales still.

The researchers are currently working to further improve the sensitivity of the equipment, which might help them probe those tiny spaces.

In a completely different context, an international team of researchers led by Professor Immanuel Bloch (LMU/MPQ) and Professor Oded Zilberberg (ETH Zürich) has now demonstrated a way to observe physical phenomena proposed to exist in higher-dimensional systems in analogous real-world experiments. Using ultracold atoms trapped in a periodically modulated two-dimensional superlattice potential, the scientists could observe a dynamical version of a novel type of quantum Hall effect that is predicted to occur in four-dimensional systems. (Nature, 4 January 2018)

"Physically, we don't have a 4D spatial system, but we can access 4D quantum Hall physics using this lower-dimensional system because the higher-dimensional system is coded in the complexity of the structure," a researcher with the US-based team, Mikael Rechtsman from Penn State University, told Ryan F. Mandelbaum at Gizmodo. "Maybe we can come up with new physics in the higher dimension and then design devices that take advantage the higher-dimensional physics in lower dimensions."

The above statements can be summed up in the following simplest model;

Let the Gravitational force between two identical particles be related to the magnetic force between them and similarly let the electric force between two particles be related to the magnetic force as;

$$\text{Gravitational force } \left(\frac{Gm^2}{R^2}\right) = \text{magnetic force (Bec)} \times \alpha_g^n \quad (103)$$

and

$$\text{Electric force } \left(\frac{e^2}{4\pi\varepsilon R^2}\right) = \text{magnetic force (Bec)} \times \alpha_e^n \quad (104)$$

Where α_e is the electromagnetic coupling constant- Fine structure constant

The magnetic flux, represented by the symbol Φ, threading some contour or loop is defined as the magnetic field **B** multiplied by the loop area, $A=\pi R^2$, i.e. $\Phi = \mathbf{B} \cdot \mathbf{A}$. Obviously, both **B** and **A** can be arbitrary and so is Φ. The inverse of the flux quantum, $1/\Phi_0$, is called the **Josephson constant**, and is denoted K_J.

However, if one deals with the superconducting loop or a hole in a bulk superconductor, it turns out that the magnetic flux threading such a hole/loop is quantized. Therefore the magnetic flux quantum from (103) and (104) will be given by,

$$\Phi_G = \pi G m^2 / ec\alpha_g^{\,n}$$

$$\Phi_E = e/4\varepsilon c\alpha_e^{\,n}$$

Such that at n=0 extra dimension,

$$\Phi_G = \pi G m^2 / ec$$

$$\Phi_E = e/4\varepsilon c$$

The above given values represent the classical flux at 3D spatial dimensions.

At n=1/2 extra dimension,

$$\Phi_G = \frac{\pi m}{e}\left(\frac{G\hbar}{c}\right)^{1/2}$$

$$\Phi_E = \left(\frac{\pi\hbar}{4\varepsilon c}\right)^{1/2}$$

These represent the quantum theory of Gravity.

At n=1 extra dimension,

$$\Phi_G = \pi\hbar/e$$

$$\Phi_E = \pi\hbar/e$$

These represent the magnetic flux quantum at the quantum scale. Also at n=1 the magnetic flux value is the same in both equations, meaning that the gravitational force becomes analogous to the electromagnetic force at n=1.

In other words, just as a 3D object casts a 2D shadow, scientists have managed to observe a 3D shadow potentially cast by a 4D object – even if we can't actually see the 4D object itself. That could unlock some new findings in the very fundamentals of science.

A Grand Unification

Physicists have argued out that the more elegant and symmetrical the theory is, the more it is beautiful. The elegancy of any physical theory is suspected at a level to which it holds well with other theories , that is ,the capability of the theory to conform with the well known laws of nature at all levels.

In this section we examine the mechanism through which quantum mechanics becomes comparable with gravity and the scale at which this occurs. At the Planck scale all interactions (the weak interaction, strong interaction and electromagnetism) are assumed to merge into a single interaction that alone occurs at very high energies of about 1TeV. The equations that do describe this phenomenon are not yet found and therefore requires one's deep effort to capture the reality of this entire puzzle.

To capture interest in these interactions we need to know first, their strength and second the range in which they occur. The strength defines the coupling constants and the range defines the attractions, on the other hand the coupling constant determines the strength of any interaction and therefore is a number in a sense that it is a dimensionless constant. A coupling constant is a very important quantity in dynamics, for example, in the motion of a large lump of magnetized iron, the magnetic forces are more important than the gravitational forces because of the relative magnitudes of the coupling constants.

The standard model is a theory of three fundamental forces - electromagnetism, weak interactions and strong interactions; however, these three forces are not tied together Howard Georgi and Sheldon Glashow discovered that the Standard Model particles can arise from a single interaction, known as a grand unified theory. Grand unified theories predict relationships between otherwise unrelated constants of nature in the Standard Model. Gauge coupling unification is the prediction from grand unified theories for the relative strengths of the electromagnetic, weak and strong forces and this prediction was verified at LEP in 1991 for supersymmetric theories.

In particle physics, supersymmetry (often abbreviated SUSY) is a novel symmetry that relates elementary particles of one spin to another particle that differs by half a unit of spin and are known as superpartners. Since the particles of the Standard Model do not have this property, supersymmetry must be a broken symmetry allowing the 'sparticles to be heavy.

One of the main motivations for SUSY comes from the quadratically divergent contributions to the Higgs mass squared. The quantum mechanical interactions of the Higgs boson causes a large renormalization of the Higgs mass and unless there is an accidental cancellation, the natural size of the Higgs mass is the highest scale possible. This problem is known as the hierarchy problem Supersymmetry reduces the size of the quantum corrections by having automatic cancellations between fermionic and bosonic Higgs interactions. If supersymmetry is restored at the weak scale, then the Higgs mass is related to supersymmetry breaking which can be induced from small non-perturbative effects explaining the vastly different scales in the weak interactions and gravitational interactions. The failure of experiments to discover either supersymmetric partners or extra spatial dimensions, as of 2006 has encouraged loop quantum gravity researchers.

The determination of the strength of the forces

We assume a model that explains everything on the length scales, the best scale so far we are familiar with is the Planck length scale, however in this model we don't associate ourselves in knowing this scale and therefore develop new scales that alone are combined together to lead to some observable phenomenon describing the forces involved in the interactions. The equation describing the model is developed and given by;

$$(v^2/c^2 + n^2\beta_{Qo}) = 8\pi\beta_{gEo} \qquad (105)$$

Where β_{Qo} is a length ratio given by l_Q/l_o, in this case $l_Q = \hbar c/W$, $\hbar$ is Dirac constant, c is the speed of light and W is the energy. Also $\beta_{gEo} = l_{gE}/l_o$ where $l_{gE} = 8\pi G M_{gE}^2/W$, G is the universal gravitational constant, M_{gE} is the mass of a particle in the combined fields given by P_{gE}/c where $P_{gE} = Gm^2ke^2/R^2c^2$, is the momentum for an elementary particle of mass m and an elementary charge e, k is the coulomb constant and R is the distance between any two particles. The equation here addresses the problems in form of length scales simply because it is at these scales that quantum mechanics seem to be comparable to gravity. The momentum P_{gE} is a momentum of a particle experiencing the strength of the electromagnetic fields and gravity. The strength is determined by a very small coupling constant as we shall see later. The smaller the distance between elementary particles, the higher the momentum and vice versa is true.

The exchange of photons between an electron and a proton in an atom is explained by Quantum Electrodynamics (QED), with a coupling constant determining the strength of the electromagnetic force. The equation of the interaction responsible for QED on the length scale, which is the Compton length, is given by the equation

$$\sum \psi^2_{fi} t_i = 2\pi \beta_{RCE} \qquad (106)$$

The expression $\sum \psi^2_{fi} t_i$ is the force changer where $\psi_{fi} = f_i R^2/ke^2$ and $t = \{f_i^2 ke^2/R^2\}/F_n^3$,

β_{RCE}, remains a constant given by l_c / l_{RE} (l_c is the Compton length $\hbar/mc$ and $l_{RE}=ke^2/mR^2c^2$). On multiplying both sides of Eqn1 by a quantity $\sum \psi^2_{fi}t_i$ we obtain,

$$(v^2/c^2 + n^2\beta_{Qo}) \sum \psi^2_{fi}t_i = 8\pi\beta_{gEo}\sum \psi^2_{fi}t_i$$

We then examine the condition for which β_{Qo} will be a maximum and minimum. It is found out from relativity that β_{Qo} is maximum when the lorentz factor $\gamma = (1-v^2/c^2)^{-1/2}$ is very small that is,

$\gamma = 1/n\sqrt{\beta_{Qo}}$ or when the velocity $v = c\sqrt{(\sum \psi^2_{fi}t_i - n^2\beta_{QO})}$

We hence obtain a general interaction equation as,

$$\sum f_y^3 \psi^2_{fi}t_i = 2\pi\beta_{RCE}\sum F_n^3 / \xi\beta_{gEo}, \quad n= 1,2,3 \qquad (107)$$

The following conditions are then taken into account

1) For $lo = l_x = mc^2/Fp$, $\beta_{gEo}=\beta_{gEx}=l_{gE}/l_x$.

2) For $l_o = l_c$, $\beta_{gEo} = \beta_{gEQ} = l_{gE}/l_c$, and

3) For $l_o = l_s = Gm/c^2$, $\beta_{gEo} = \beta_{gEs} = l_{gE}/l_s$, which gives

$$\sum f_y{}^3 \psi^2{}_{fi\,ti} = F_1{}^3 + F_2{}^3 + F_3{}^3 = 2\pi\beta_{RCE}\left(F_p{}^3/8\pi\beta_{gEx} + F_p{}^3/256\pi^3\beta_{gEQ} + 2F_B{}^3/\pi\beta_{gEo}\right) \quad (108)$$

Where $F_p = c^4/G$ is the Planck unit force and $F_B{}^3 = m^2 c^3/\hbar$ is the force required for strong and weak interactions to take place. Again setting a condition,

For for $l_o = l_z = ke^2/mc^2$, $\beta_{gEo} = \beta_{gEz} = l_{gE}/l_z$.

$$\sum f_y{}^3 \psi^2{}_{fi\,ti} = F_4{}^3 = 2\pi\beta_{RCE}\left(F_z{}^3/32\pi^3\beta_{gEx}\right)$$

Where $F_z{}^3 = m^2 c^4/ke^2$,

Also for $l_o = l_N = \hbar^2 m^3 G^2/k^3 e^6$, $\beta_{gEo} = \beta_{gEN} = l_{gE}/l_N$, we obtain,

$$\sum f_y{}^3 \psi^2_{fi} t_i = F_5{}^3 = 2\pi \beta_{RCE} (F_z{}^3/2\pi^3 \beta_{gEN}) \qquad (109)$$

Measuring the value of the strong, weak and electromagnetic coupling constants gives us away through which we can determine supersymmetric levels. From supersymmetry and grand unification of elementary particles the couplings agree to 1%. The relationships of the sum of the cubes of the forces to each individual cube of the force, and that of the sum of the square of masses with each known mass squared casts much information about the masses and couplings of the supersymmetric particles as shown below, when Eqn4 is divided through respectively by the cubes of the forces $F_1{}^3$, $F_2{}^3$ and $F_3{}^3$ the following equations are obtained,

$$\sum F_n{}^3/F_1{}^3 = 1 + 16\alpha_g{}^3 + 1/32\pi^2\alpha_g \qquad (110)$$

$$\sum F_n{}^3/F_2{}^3 = 1 + 32\pi^2\alpha_g + 512\pi^2\alpha_g{}^4 \qquad (111)$$

$$\sum F_n{}^3/F_3{}^3 = 1 + 1/6\alpha_g{}^3 + 1/512\pi^2\alpha_g{}^4 \qquad (112)$$

$$\sum F_n{}^3/F_4{}^3 = 1 + \beta^2(4\pi^2 + 1/8\alpha_g) + 64\pi^2\alpha_s{}^2\alpha_g \qquad (113)$$

Where $\beta = ke^2/Gm^2$ is the ratio of the fine structure constant α_s to the gravitational coupling constant α_g, given respectively as $\alpha_g = Gm^2/\hbar c$ and $\alpha_s = ke^2/\hbar c$.

Now equating $F_4 = F_5$, $F_5 = F_3$, $F_5 = F_1$ we obtain; m_1, m_2, m_3 and m_4 respectively, Adding the squares of the masses we obtain,

$$\sum m_n^2 = m_1^2 + m_2^2 + m_3^2 + m_4^2 \quad (1114)$$

Which gives the sum per unit mass as,

$$\sum m_n^2/m_1^2 = 1 + 16\pi^2\alpha_s^4 + (8\pi/\alpha_s)^{1/2} + 4/(128\,\alpha_s^4)^{1/5} \quad (115)$$

$$\sum m_n^2/m_2^2 = 1 + 1/16\pi^2\alpha_s^4 + (1/8\pi\,\alpha_s^9)^{1/2} + (1/4\pi^2)(128\,\alpha_s^{24}) \quad (1116)$$

The equations generated so far give a basis for the nature and type of supersymmetry exhibited by a particle experiencing forces at both the Planck and grand unified scales. It is thus shown here that the electromagnetic coupling constant is a result of mathematically summing the squares of the masses generated and then dividing through by the square of the mass in the summation while the gravitational coupling constant is the result of summing the cubes of the forces and then dividing through by the cube of the force in the sum. This idea at its best is taken to be the basis for symmetric theories as we shall see in the results obtained.

Results

The unification of coupling calculations

At equal forces that is $F_1 = F_2 = F_3 = F_p$ the mass $M_p = (\hbar c /8\pi G)^{1/2} = 2.1765 \times 10^{-8}$ kg, is obtained which is the Planck mass for which the Schwarzschild radius is equal to the Compton length divided by π. When Eq4 is divided through by $F_1{}^6$ and $F_2{}^6$ we obtain equations of the form;

$$\sum F_n{}^3/F_1{}^6 = \Omega/F_p{}^3 \quad (117)$$

$$\sum F_n{}^3/F_2{}^6 = \text{€}/F_p{}^3 \quad (118)$$

Where, $\Omega = 4m^2/m_p{}^2 + 1/\pi + m^8/32\pi^2 m_p{}^8$ and $\text{€} = m^6/8\pi m_p{}^6 + 16\pi m^4/m_p{}^4 + 2m^{12}/\pi m_p{}^{12}$

The mass relations equations obtained above indicate the scale at which gravity may be strong and weak. Obtaining these results on the Planck force and mass scale is evidence for the existence of the theory of quantum gravity. The values Ω and € represent a series equation defined by increasing powers in the mass ratio (m/m_p). The mass m is assigned to any particle and the mass m_p is assigned to the Planck scale defining quantum gravity.

The unit of energy is $M_P c^2$; the unit of electric charge is $\sqrt{hc/k}$, where k is coulomb constant and so forth. On the other hand, one cannot form a pure number from these three physical constants. Thus one might hope that in a physical theory where ℏ, c, and G were all profoundly incorporated, all physical quantities could be expressed in natural units as pure numbers. Within its domain, this paper has achieved it for example, imagining that there were just two quark species with vanishing masses. Then from the two integers 3 (colors) and 2 (flavors), ℏ, and c (without mass parameters), the spectrum of hadrons with mass ratios and other properties close to those observed in reality, emerges by through calculation (Ω and ϵ) as indicated from Eqn117 and Eq118 shown above. The overall unit of mass is indeterminate, but this ambiguity has no significance within the theory itself. The results obtained show an ideal Planckian theory that alone does not contain any pure numbers as parameters. Thus, for example, the value $m_e/m_p = 10^{-22}$ of the electron mass in Planck units is obtained from a dynamical calculation. This ideal might be overly ambitious, yet it seems reasonable to hope that significant constraints among physical observables will emerge from the inner requirements of a quantum theory which consistently incorporates gravity. The model therefore provides; first, the unification of couplings calculation. second, it points to a symmetry breaking scale remarkably close to the Planck scale (though apparently smaller by 10^{-2} to 10^{-3}), so there are pure numbers with much more 'reasonable' values than 10^{-22} to shoot for. Third, it shows quite concretely how very large scale factors can be controlled by modest ratios of coupling strength, due to the logarithmic nature of the running of couplings (so that 10^{-22} may not be so 'unreasonable' after all).

While the above result is based on the study of the strength of the gravitational force, we now look for ways in which we can examine the strength of the electromagnetic force depending on the mass. This is done by dividing the sum of the squares of the masses (Eqn114) by the fourth power of the individual masses hence,

$$\sum m_n^2 / m_2^4 = \omega / m_G^2 \quad (119)$$

$$\sum m_n^2 / m_E^4 = \lambda / m_p^2. \quad (120)$$

Where $\omega = 1/\ 16\pi^4 \alpha_s^9 + 1\ /4\pi^2 \alpha_s^5 + 1/\ (512\pi^8\ \alpha_s^{19}\ )^{1/2}$, $\lambda = 128\pi^3 \alpha_s^6 + 8192\pi^5 \alpha_s^{10} + 128\sqrt{\pi^3} \alpha_s^{11} + 128(\pi^{15} \alpha_s^{26})^{1/5}$

$m_E = (1/\ 8\pi\ Ke^2)(\hbar^3 c^3/G)^{1/2}$ is the mass obtained when $F_4^3 = F_3^3$, and $M_G = (Ke^2/G)^{1/2}$ is the mass obtained when the electromagnetic force is equal to the gravitational force.

It can now be theorized that the strength of the electromagnetic force is determined by Eqn119 and 120 at which a series power equation in the fine structure constant defined by ω and λ is a constant.

The length scales at which the masses predicted by the standard model survive

The mass of the W and Z bosons (M_W, M_Z), Higgs particle (M_H) and the mass scale at the grand unification (M_{GUT}) are generated. We multiply a coupling constant μ with the force F_3^3, of which we equate to F_4^3 that is;

$$\mu F_3^3 = F_4^3$$

From which

$$\mu = R_B^2 / R_o^2,$$

Where R_o is the length scale determined experimentally and $R_B = (8\pi G k e^2/c^4)^{1/2} = 6.9101 \times 10^{-36}$ m, which is greater than the Planck length.

So the equation that produces the different masses at R_o will be given by the square of the mass as,

$$M^2 = m_p^2 / 8\pi\mu \, \alpha_s^{\,2}$$

Where $\alpha_s = 1/137$, is the electromagnetic coupling constant.

To obtain the masses, we need to find the length R_o, theoretically we develop the lengths given by; 1.03741×10^{-39} m, 8.3182×10^{-54}m, 9.4334×10^{-54}m, and 1.2345×10^{-53}m.

Following the given lengths we respectively obtain the masses;

$M_{GUT} = 10^{16}$ GeV, $M_W = 80.18$ GeV, $M_Z = 90.82$ GeV, and $M_H = 119$ GeV respectively.

But at $R_B = R_o$, the mass $M_B = 6.661 \times 10^{19}$ GeV is obtained. And at $R_o = 2.529 \times 10^{-37}$m, the Planck mass is obtained (that is $M = m_p$). Therefore it is found out that the W and Z boson particles survive in length of 10^{-54}m. The Higgs particle survives to a length greater than that of the W boson $\geq 10^{-53}$m. And finally particles at the grand unified scale will survive at 10^{-39} m.

The big bang acceleration and proton decay

For proton decay the intensity P is used such that at Schwarzschild radius R and Planck mass scale m_p the life time of the proton as explained by SUSY is seen to agree so well with the

$$T(time) = \alpha^2 \, m_p^5 \, R \, / \, 4096 \, \pi^3 m_k^4 \, \hbar$$

Such that at $m_k = 7.96 \times 10^{-29}$ GeV, $T = 10^{35}$ yrs.

We have obtained the lifetime of protons and the mass of a particle produced during the decay process. The mass of the particle obtained is very small and can therefore be taken to be a neutrino.

The force F_3 can be expressed in the form,

$$F_3 = a_3(m_3^5 / 16\pi^2 m_p^2)^{1/3}$$

Where a_3 is the acceleration, this acceleration at a Planck scale will be given by

$$a_3 = (c^{11} / \hbar G^2 m)^{1/3} = 2.4772 \times 10^{52} \text{m/s}^2$$

This is quite a very large acceleration and therefore defined as the acceleration of particles during the early formation of the universe.

The results obtained describe super symmetry which is a theory required for the unification of everything we know about the physical world into a theory of everything. Significantly a larger enterprise of the theory is to produce a theory of quantum gravity which is required for the unification of general relativity with the standard model, which explains the other three basic forces in physics (electromagnetism, the strong interaction, and the weak interaction), and provides a palette of fundamental particles upon which all four forces act. Theoretically the results obtained (Eqn115 and Eqn116) show a huge correction to the particles' masses, which without fine-tuning will make them much larger than they are in nature. The problem of the unification of the weak interactions, the strong interactions and electromagnetism is solved mathematically, through the comparisons of the cube of the forces in a ratio that generates the gravitational coupling constant power equation.

The Planck mass is the mass of a black hole whose Schwarzschild radius multiplied by π equals its Compton wavelength. The radius of such a black hole is roughly the Planck length, which is believed to be the length scale at which both general relativity and quantum mechanics simultaneously become important. In accordance with the results obtained it is seen that the Planck mass is the mass at which the four forces (F_1, F_2, F_3 and F_p) are equal, the forces are then taken to be related to the origin of the universe simply because at those high energies that formed the dense soup of the universe the forces were equal and the masses probing the Planck mass scale that is black holes were produced, hence those four forces a significant in that they play a crucial role in the formation of black holes. The intensity P on the other hand explains a phenomenon that occurs at the cosmic scale, for example it explains the nature of Black holes and the age of the universe. The acceleration obtained is so large that it is the acceleration that the universe had at the instant after the big bang. Obtaining this acceleration is the possibility of studying the rate of expansion of the universe at large, the accelerating universe is therefore the observation that the universe appears to be expanding at an accelerated rate.

At the Planck scale the descriptions of subatomic particle interactions in terms of quantum field theory breaks down. Also at the same scale, the strength of gravity is expected to become comparable to the other forces, mathematically all the fundamental forces are unified at that scale. The results obtained explain both the weak and strong interactions that at a length between 10^{-37}m and 10^{-35} the Planck scale is attained also at lengths 10^{-39}m , the grand unified scale becomes relevant , but for lengths 10^{-53}m and 10^{-54}m, the standard model holds on well. We have therefore attained a unification that increases from about 10^{-59}m (standard model) to 10^{-35}m (quantum gravity).The paper there fore gives out the relationship between elementary particle physics and astrophysics at a large scale.

Basing on the results obtained, it is now clearly justified that gravity can be integrated with quantum mechanics at the Planck scale. And therefore the success of the "standard model" which includes both the electroweak theory and quantum chromodynamics can now be regarded as successful in providing accurate descriptions of the fundamental particles and their interactions.

A Unified Bohr and Quantum Gravity Theory

In the early 20th century, Ernest Rutherford experiments established that atoms consisted of a diffuse cloud of negatively charged electrons surrounding a small, dense, positively charged nucleus. Given his experimental data, it was quite natural for Rutherford to consider a planetary model for the atom, the Rutherford model of 1911, with electrons orbiting a sun-like nucleus. This model was a difficulty. The laws of classical mechanics predict that the electron will release electromagnetic radiation as it orbits a nucleus. Because the electron would be losing energy, it would gradually spiral inwards and collapse into the nucleus. This was a disaster, because it predicted that all matter was unstable.

To overcome this difficulty, Niels Bohr proposed, in 1913, what is now called the Bohr model of the H atom. The model's key success laid in explaining the Rydberg formula for the spectral emission lines of atomic hydrogen. Not only did the Bohr model explain the reason for the structure of the Rydberg formula, but it provided a justification for its empirical results in terms of fundamental physical constants.

This section looks at the model in a very different way than that of Bohr. The fact that all accelerated particles do emit electromagnetic radiations is taken into account and therefore the acceptance for the unstableness of all matter is considered in due respect. In fact Bohr's ideas never required classical mechanics simply because it could not conform to the experimental observations of the spectrum of the Hydrogen atom that were obtained by Rydberg using his formula.

To merge gravity with Planck's quantum theory by then was also a problem at hand and therefore Bohr had to forego the problem by introducing in his theory adhoc postulates, and this could have been the reason why Einstein found problems in merging gravity with electromagnetism in what is called "The Grand unified field theory", of which he had to question the problem with the quantum theory and therefore request for a complete quantum theory. From Bohr's model many theories have been formed each building from the ideas of the model, but a certain point is reached where the theories can not conform well to the known laws of nature and therefore regarded as failures, which of course in their judgments is true. The problem is seen to come from exactly the roots of quantum mechanics.

The aim of this section is therefore to produce a generalized theory of atomic structure that incorporates in it gravity and quantum mechanics and thus predict the properties of the universe at the Planck era.

The Hydrogen atom exists in certain stationary states of discrete energies. The acceleration due to gravity of an electron in orbit around the nucleus will cause the atom to emit radiations (radiate energy) and thus make the atom unstable. The acceleration (g) falls off with time t provided the radius of orbit of the electron R is a constant thus the acceleration due to gravity is given by;

$$g = R/\Delta t^2 \qquad (121)$$

The rate of change of energy P radiated as a result of the above acceleration will depend on the constants c (speed of light) and G (universal gravitational constant), hence;

$$P = c^5/G \qquad (122)$$

The power and time must be re- quantized in units of $\hbar = h/2\pi$ where h is Planck constant, hence

$$P\Delta t^2 = n^2\hbar \qquad (123)$$

Where n= 1,2,3…….. is the principle quantum number.

But the total energy of the atom in the various energy states is $W = -ke^2/R$ where k is the Coulomb constant and e is the elementary charge. Since Δt^2 is known from Eqn1 and P from Eqn2 then using Eqn3 the radius is given by

$$R = n^2 Gg\hbar/c^5 \qquad (124)$$

From which the total energy is given by,

$$W = -ke^2c^5/n^2Gg\hbar \qquad (125)$$

From the Bohr-Einstein frequency (f) condition, applied to a transition from a level with $n = n_i$ to a level with $n = n_f$, The energy of a photon emitted by a hydrogen atom is given by the difference of two hydrogen energy levels

$$hf = E_i - E_f$$

Finally we get since frequency $f = c/\lambda$, where λ is the wavelength

$$1/\lambda = [kc^2c^4/2\pi G\hbar^2][1/g][1/n_f{}^2 - 1/n_i{}^2] \qquad (126)$$

The equation obtained above shows somehow a great significance of gravity in the quantum theory. So far it states that regardless of the levels in the transitions of an atom the acceleration due to gravity of the particles in the atom do greatly affect the nature of its spectrum.

The quantity $[ke^2c^4/ 2\pi G\hbar^2]$ is the inverse of the square of time t and therefore; $1/t^2 = [ke^2c^4/ 2\pi G\hbar^2]$, from which the time is obtained as $t = 1.58873 \times 10^{-42}$s. This is the earliest period of time in the history of the universe.

Comparing Eq126 with Bohr's model, here we shall equate the Rydberg constant $[k^2e^4m/4\pi c\hbar^3]$, where m is the mass of the particle, to the constant $[ke^2c^4/ 2\pi G\hbar^2][1/g]$. Doing this generates an acceleration given by $g_a = [2\hbar c^5/ke^2 Gm]$, then at the Planck epoch when $m = \sqrt{\left(\frac{\hbar c}{G}\right)}$ the acceleration reduces to $g = \frac{8\pi \varepsilon_0}{e^2}\sqrt{\left(\frac{\hbar c^7}{G}\right)}$. Then At the Schwarz child's radius $R = Gm/c^2$ the acceleration is $g_b = c^4/Gm$ which gives an equation for the spectrum as $1/\lambda = [/ 2\pi a_o][1/ n_f^2 - 1/ n_i^2]$ where a_o is the first Bohr radius $[\hbar^2/ mke^2] = 5.28 \times 10^{-11}$m.

The interesting part of it is that the ratio $g_b / g_a = [ke^2/2\hbar c]$ is the fine structure constant.

Everything

Quantum gravity is the field of theoretical physics that tries to unify quantum mechanics with general relativity. Quantum mechanics describes the three fundamental forces of nature while general relativity is a theory of the fourth fundamental force: gravity. The goal everyone is waiting for to emerge from this unification is a "theory of everything", or "Grand Unified Theory" (GUT). In 1986, Abhay Ashtekar reformulated Einstein's field equations of general relativity using what have come to be known as Ashtekar variables, a particular flavor of Einstein-Cartan theory with a complex connection. He was able to quantize gravity using gauge field theory. In the Ashtekar formulation, the fundamental objects are a rule for parallel transport and a coordinate frame known as a vierbein at each point. Because the Ashtekar formulation was background-independent, it was possible to use Wilson loops as the basis for a nonperturbative quantization of gravity. Explicit (spatial) diffeomorphism invariance of the vacuum state plays an essential role in the regularization of the Wilson loop states.

Around 1990, Carlo Rovelli and Lee Smolin obtained an explicit basis of states of quantum geometry, which turned out to be labelled by Penrose's spin networks. In this context, spin networks arose as a generalization of Wilson loops necessary to deal with mutually intersecting loops. Mathematically, spin networks are related to group representation theory and can be used to construct <u>knot invariants</u> such as the Jones Polynomial.

The need for this chapter is to understand those problems involving the combination of very large mass or energy and very small dimensions of space, such as the behavior of black holes, and the origin of the universe.

The formula for the quantization of quantum gravity

The model is based on separating the gravitational field into the sum of two components; that is the background and the quantum field. The background left is one for all our calculations. But because loop gravity ignores the back ground space as a lost entity that does not occur in space, there fore the need to reconstruct quantum field theory from scratch without a background space is taken into account. I therefore suggest that the calculation should be performed by summing all possible space-times.

Quantum field theory depends on particle fields embedded in the flat space-time of special relativity. General relativity models gravity as a curvature within <u>space-time</u> that changes as a gravitational mass (m) moves. Assuming a spherical symmetric object that space time is of dimensions increasing from 1, 2, 3, 4...N, where N is the nth term of the dimensions. To quantize space and time is to create a space in which all of physics is quantized. The nature of the curved space surface is described by increasing powers in the Schwarzschild radius $R_s = Gm/c^2$, Hence describing the dimensions of space. Quantum mechanics explains the existence of discrete energy states in an atom, in away that the angular momentum of the atom must be quantized, which is also the case for quantum gravity. The equation for the quantization of the loop quantum gravity can then be written as,

$$\eta R_s + \beta R_s^2 + \mu R_s^4 + \ldots\ldots\ldots + \delta R_s^N = n\hbar \qquad [127]$$

Where $\eta = \sqrt{Beh}$, is the momentum of a particle probing another form of quantum mechanics, $\hbar = h/2\pi$, where h is Planck constant, $\beta = 8\pi Be$, e is the elementary charge, B is the magnetic field and finally $\mu = 256\pi^3 P/c^2$, where P is the intensity and c is the constant speed of light.

The energy equation

What changes is the form of the equation the rest remaining constant. The principle behind this is that eqn1 can be changed to any form simply for purposes of calculating complex phenomenon. The energy to which we are concerned here is expressed as a general expression describing the energy scales forming smaller and larger matter entities in the universe. The energy will thus be given by;

$$\eta c + \beta c R_s + \mu c R_s^3 + \ldots\ldots\ldots + \delta c R_s^{N-1} = n\hbar c/R_s$$
[128]

Note: the background space described by the Schwarzschild radius has changed, thus the above equation in any case can be used to calculate the basic properties of Black holes. Remember the Schwarzschild radius is the radius for a given mass where, if that mass could be compressed to fit within that radius, no known force or degeneracy pressure could stop it from continuing to collapse into a gravitational singularity.

The mass equation

Having explored the energy scale we now form general equation that describes well the mass scale. This is also done the same way as eqn128 and therefore generate,

$$\eta/c + \beta R_s/c + \mu R_s^3/c + \ldots + \delta R_s^{N-1}/c = n\hbar/cR_s$$
[129]

The maximal magnetic field

Assuming that the energy $W = \beta cR_s$, from eqn128 is equal to the energy $W = mc^2$, we hence obtain the magnetic field as, $B = c^3/8\pi Ge = 1.0054 \times 10^{53}$ N/Am. using this magnetic field in the energy equation, $W = \eta c$ we get the energy in the form $W = (c^2/2)\sqrt{\hbar c/G}$ where the quantity $\sqrt{\hbar c/G}$ is the Planck mass M_p at an energy of 6.119×10^{18} Ge

Time taken by a black hole to evaporate and its entropy

The energy required here is given in Eqn128, it is at this, that the intensity $P = W/A\Delta t$, (where A is the area and t is the time) is used. We take the energy $W = \mu c R_s^3$ (from Eq128) as our interest from which we obtain the time as $\Delta t = 256\pi^3 R_s^3/Ac$. But with black holes the area will become exactly equal to the square of the Planck length as $A \sim L_p^2 = \hbar G/8\pi c^3$ hence the change in time is given by $\Delta t = 63500.86\pi G^3 m^3/\hbar c^4$.

For entropy we set the energy to kT, where k is Stefan's-Boltzmann's constant and T is the temperature of the body. Now for $kT = \mu c R_s^3$, since Δt is known the entropy is thus given by $S = W/T = 78.96 Ak c^3/\pi \hbar G \sim A/4$. In conclusion we state that the entropy of a black hole is proportional to the area of the event horizon.

The quantum Hall Effect

For this effect the momentum η is used. From Eqn128 we set, $\eta c = n\hbar / R_s$ which gives the magnetic flux as $4\pi R_s^2 B = nh/e$, from which the resistance is given by $\zeta = 4\pi R_s^2 B /e = nh/e^2$. for n= 1,2,3,4 the resistance is of a value 25833.8Ω

Maximum Intensity

Using eqn129 in this case, since B is known and P got from $\mu R_s^4 = n\hbar$; as $P = \hbar c^2/256\pi^3 R_s^4$, we hence obtain, $M_p/2 + m + M_p/m = M_p/m$, which gives $M_p + 2m = 0$, and for identical mass $M = 0$, which is true. The intensity at the planck length that is for $R_s = L_p$ is $P = c^8/\pi\hbar G^2$

The Theory of Light

Basing our study on the electric currents generated whenever there is a changing magnetic field (B) and a changing electric field (E) in the electromagnetic wave we can construct a complete theory for the electromagnetic radiations. The theory is created using the symmetry between a long wire placed in the electromagnetic fields which induce vibrating electrons that carry current in the wire and the electromagnetic wave which constitute changing electric and magnetic fields that create vibrating photons in the wave. Therefore a wire is to a wave what a vibrating electron is to a vibrating photon in the wire and a wave respectively. The aim of the paper is to give a clear description of the theory of electromagnetic radiations (light). The goal of the paper on the other hand is to show that the wave-particle descriptions of reality can be applied to any physical situation simultaneously. The objective of the paper is to show that the Photoelectric Effect and the Compton Effect can both be explained by the wave model and the particle model at the same time.

Consider a long wire connected to an ammeter and strong electric and magnetic fields produced in a vacuum. Let us assume that whenever a wire is brought in vicinity of a changing electric field, electrons of mass (m) are set into motion in the wire and then an ammeter deflects, recording a current (i_E). The current in the wire due to a changing electric field should be given by

$$i_E = \frac{j\varepsilon_o}{2\pi m}E \qquad (130)$$

Where (ε_o) is the permittivity of free space and (j) is the constant of action in SI units Js. therefore the current is quantized and depends on both the electric field and the mass of an electron.

When the wire is brought into the magnetic field, vibrating electrons at a frequency of oscillation (f) are set in motion at a speed (v) through the wire generating a current given by

$$i_B = \frac{v}{2\pi \mu_o f}B \qquad (131)$$

Where (μ_o) is the permeability of free space.

Assuming that the ammeter records different values of (i_E) and (i_B), what will be the change in the current values recorded at the ammeter? Subtracting equation (130) from equation (131) we have

$$\Delta I = (i_E - i_B) = \left(\frac{j\varepsilon_o}{2\pi m}E - \frac{v}{2\pi \mu_o f}B\right) \qquad (132)$$

This is the change in the currents due to changing magnetic and electric fields. Assuming that there is no change in the current, meaning that the current values for i_E are equal to those of i_B (i.e $\Delta I = 0$). This will imply that the magnetic field strength was equal to the electric field strength at one point in both experiments. In terms of electromagnetic radiations in the vacuum, assuming that a wire carrying current is replaced by a wave and electrons are replaced by photons. The wire replaced by a wave is made up of vibrating electric and magnetic fields at a given frequency making an electromagnetic wave. The electrons replaced with photons will represent the particle properties of the electromagnetic wave (light) with associated mass and speed (v).

The symmetry here is between the long wire and the wave, the electrons and the Photons. The electric and magnetic fields brought in vicinity of the wire and the number of oscillations per second of the electron in the wire is what leads to an electromagnetic wave. The electrons with a given mass and moving at a given speed is what constitute a photon. Then at $\Delta I = 0$, we have on arranging,

$$\frac{jf}{mv} = \frac{1}{2\pi\mu_0\varepsilon_0}\frac{B}{E} \qquad (133)$$

This means that at $\Delta l = 0$, either a changing magnetic field or a changing electric field produces a current. Then it should be true that a changing magnetic field produces an electric field just as a changing electric field produces a magnetic field. This process in the electromagnetic wave continues indefinitely. The electromagnetic wave will move at a constant speed (c), since for electromagnetic waves, $\frac{E}{B} = c$, and for a photon $\frac{jf}{mv} = c$ where $j = 6.63 \times 10^{-34}$ Js (also called the Planck constant after Max Planck) and mv is the photon momentum. Implying that the photon energy is related to the frequency of the electromagnetic wave by (jf). Then the electromagnetic wave will move at a constant speed given as, since by symmetry $\frac{E}{B} = \frac{jf}{mv} = c$

$$c = \frac{1}{\sqrt{\varepsilon_o \mu_o}} = 2.99792458 \times 10^8 \frac{m}{s}$$

Where $\varepsilon_o = 8.85418782 \times 10^{-12} \frac{c^2}{Nm^2}$ and $\mu_o = 1.26 \times 10^{-6} \frac{Ns^2}{c^2}$

We have therefore deduced based on the symmetry between a current (electron) carrying wire in the electromagnetic field and the photons in electromagnetic waves that an electromagnetic wave moves at a constant speed of light. It is also true from the deductions that light is indeed made up of particles of light called photons and vibrating electric and magnetic fields. The deduction would not be possible if the wave and particle descriptions of the situations had not been applied simultaneously (into what is called "the wave-particle duality).

Photoelectric effect

Unexpectedly enough the photoelectric effect can also be explained by Equation (3), on arranging

$$\frac{2\pi m f}{\varepsilon_0 E} \Delta l = jf - \frac{mv}{2\pi \mu_0 \varepsilon_0} \frac{B}{E}$$

Then the total energy of the particle of light (Photon) is then given by

$$jf = \frac{2\pi m f}{\varepsilon_0 E} \Delta l + \frac{mv}{2\pi \mu_0 \varepsilon_0} \frac{B}{E} \qquad (134)$$

It is therefore true that the photoelectric effect can be explained when both the particle and wave models of reality are applied in the experiment at the same time (simultaneously). The work function from Einstein's photoelectric equation (A. Einstein, 1905) will here be replaced by $\frac{2\pi m f}{\varepsilon_0 E}\Delta l$ while the kinetic energy of the electrons at the surface of the metal will be given by $\frac{mv}{2\pi \mu_0 \varepsilon_0} \frac{B}{E}$. Equation (134) reduces to Einstein's Photoelectric effect when, the speed of the electron is

$v = \frac{1}{\pi\mu_0\varepsilon_0}\frac{B}{E}$ and the change in current for a complete circuit is $\Delta I = \frac{j\varepsilon_0 E}{2\pi m}$.

Compton Effect

The validity of the Compton Effect can also be deduced from Equation (132). The current can be taken as the product of the frequency (f) of radiations and the charge (q) on the particle.

Then the current due to the electric field is $i_E = qf_1$ and that due to the magnetic field is $i_B = qf_2$. In the case of the Compton Effect, q is the charge on the free electron while f_1 and f_2 are the frequencies of the incoming photon and outgoing photon after collision with the free electron respectively. Then equation (132) can be written as

$$f_1 - f_2 = \frac{1}{q}\left(\frac{js_o}{2\pi m}E - \frac{v}{2\pi \mu_o f}B\right) \qquad (135)$$

Since photons move with the speed of light(c) then their frequencies is related to their speed and wavelength by $f = \frac{c}{\lambda}$, then we have

$$\frac{1}{\lambda_1} - \frac{1}{\lambda_2} = \frac{1}{qc}\left(\frac{js_o}{2\pi m}E - \frac{v}{2\pi \mu_o f}B\right)$$

On arranging to include the charge density of the free electron for electric field lines in an area of $\frac{\lambda_1 \lambda_2}{2\pi}$, we obtain

$$\frac{2\pi q}{\lambda_1 \lambda_2}(\lambda_2 - \lambda_1) = \frac{j}{mc}\left(\varepsilon_o E - \frac{mv}{\mu_o jf}B\right)$$

Where (mc) is the momentum of an electron treated relativistic ally, on letting the charge density $= \frac{2\pi q}{\lambda_1 \lambda_2} = \varepsilon_o E$, we deduce the change in the wave length of the incoming photon and outgoing photon after collision with the free electron as

$$\Delta\lambda = (\lambda_2 - \lambda_1) = \frac{j}{mc}\left(1 - \frac{mv}{\rho\mu_o jf}B\right)$$

Since $\rho = \varepsilon_o E$, we then have

$$\Delta\lambda = (\lambda_2 - \lambda_1) = \frac{j}{mc}\left(1 - \frac{\frac{mvB}{\varepsilon_o \mu_o E}}{jf}\right)$$

Since jf is the energy carried by the photon, and then also $\frac{mvB}{\varepsilon_o \mu_o E}$ is the energy carried by the free electron.

272

Treating the electron relativistically such that for electromagnetic waves moving at a speed (v) relative to the electron moving at a speed of light $c = \frac{1}{\sqrt{\varepsilon_0\mu_0}}$, the electric field in the wave will be related to the magnetic field by $Bv = E$. then the energy carried by an electron can be given by mc^2. Then the angle at which the photon is scattered after collision with the free electron will be given by

$$\theta = \cos^{-1}\left(\frac{mvB}{\varepsilon_0\mu_0 E}\right) / jf \qquad (136)$$

Where mv is the momentum of the photon in the electromagnetic wave consisting of a changing electric field E and magnetic field B both moving at a constant speed of light $c = \frac{1}{\sqrt{\varepsilon_0\mu_0}}$. Treating the electron relativistically we have

$$\theta = \cos^{-1}\frac{mc^2}{jf}$$

When the energy carried by the photon is equal to the energy possessed by the electron then $\theta = 0$, meaning that

273

there is or there is no scattering and whatsoever there is no increase in photon wavelength hence $\Delta\lambda = 0$.

A complete theory of light can't fail to explain the structure of an atom. I therefore take a complete discussion of what goes on inside an atom only with the help of Bohr's energy levels which he derived using classical mechanics and quantum theory. Let $\Delta f = f_1 - f_2$ be an increase in the frequency of the electromagnetic radiations emitted from an atom. Then squaring both sides of equation (135) and arranging will give

$$4\pi^2 \Delta f^{2^1} = \frac{1}{m^2 q^2}\left(j\varepsilon_o E - \frac{mv}{\mu_o f}B\right)^2$$

$$4\pi^2 m^2 q^2 \Delta f^2 = j^2\varepsilon_o^2 E^2 - 2\frac{j\varepsilon_o EBmv}{\mu_o f} + \frac{B^2 m^2 v^2}{\mu_o^2 f^2}$$

Dividing through by $64\pi^4 j^2 \varepsilon_o^2$ and multiplying through by q^2 gives the energy of the atom as on arranging

$$\frac{mq^4}{16\pi^2 n^4 j^2 \varepsilon_o^2} = \frac{1}{64\pi^4 m\Delta f^2}\left((Eq)^2 - 2\frac{m(Eq)(Bqv)}{\mu_o \varepsilon_o (jf)} + \frac{(Bqv)^2 m^2}{\mu_o^2 \varepsilon_o^2 (jf)^2}\right)$$

The energy of the n-th level is since the reduced Planck constant is

$$n\hbar = \frac{nj}{2\pi}$$

$$\frac{mq^4}{32\pi^2\pi^4 n^2\hbar^2\varepsilon_0^2} = \frac{1}{32\pi^2 n^2 m\Delta f^2}\left((Eq)^2 - 2\frac{m(Eq)(Bqv)}{\mu_0\varepsilon_0(jf)} + \frac{(Bqv)^2 m^2}{\mu_0^2\varepsilon_0^2(jf)^2}\right)$$

The expression on the left hand side of the equation is the quantized energy of an atom (Niels Bohr, 1913) while the right hand side of the equation represents the energy of the atom in terms of the forces associated with it. In the equation we let $H_e = Eq$ be the electric force for a particle moving in the electric field and $H_b = Bqv$, the magnetic force on a particle with charge q moving in the magnetic field. Since the speed of light is $= \frac{1}{\sqrt{\varepsilon_0\mu_0}}$, then the quantized energy can be given as

$$W_n = \frac{1}{32\pi^2 n^2 m\Delta f^2}\left(H_e^2 - 2\frac{H_e H_b mc^2}{jf} + \frac{H_b^2(mc^2)^2}{(jf)^2}\right)$$

Then on arranging we obtain

$$W_n = \frac{1}{32\pi^2 n^2 m\Delta f^2}\left(H_e - \frac{mc^2}{jf}H_b\right)^2 \qquad (137)$$

When the energy of an electron moving at a speed of light in atom is equal to the energy of the emitted photon, then

$$W_n = \frac{1}{32\pi^2 n^2 m \Delta f^2}(H_e - H_b)^2 = \frac{1}{32\pi^2 mn^2}\left(\frac{\Delta H}{\Delta f}\right)^2$$
(138)

Where $\Delta H = H_e - H_b$ is the difference or change between the electric force and the magnetic force in an atom, when the two forces balance (i.e. $H_e = H_b$), then $W_n = 0$ meaning that the total energy of an atom will cease to exist.

Therefore the total energy of an atom increases with the square of the change in the electric and magnetic forces which govern an electron but falls off as the square of the change in the frequency of the radiation emitted by it.

From equation (137) the ratio of the energy of an electron to that of the photon $\frac{mc^2}{jf}$, is the limit at which if the energies are not equal you will not get a change in the electric and magnetic forces. Treating the ratio as a number $\tau = \frac{mc^2}{jf}$, we get from equation (8)

$$W_n = \frac{1}{32\pi^2 mn^2} \left(\frac{H_e - \tau H_b}{f_1 - f_2}\right)^2 \qquad (139)$$

When $\tau = 0$, it means that the relativistic energy (mc^2) of an electron in an atom is zero, and that the total energy of an atom only increases with the electric force on the electron. The relationship (equation 139) is a complete expression for the laws according to which, by the theory here advanced, the structure of an atom should be viewed.

In conclusion, a complete theory of light is only possible if both the wave and particle descriptions of reality are applied to the physical situation at the same time. In discussing Young's double slit experiment for example we should be able with the formulas given above to treat the electromagnetic radiations on both a wave and particle model.

Making Sense with Semi-Classical Gravity

For the past thirteen years, I have been working on the most important theories of physics from scratch without employing the methods of general relativity and quantum field theories and I have come up with promising results. I have deduced the Black hole thermodynamics from first principles, I have deduced the Wiedmann Franz law from scratch, the Stefan Boltzmann power law, The result for the earliest period of time in the history of the Universe, I have related the Chandrasker theory of white dwarfs with the Bohr theory of the Hydrogen atom-the results are suprising, the rest is history. This book gives a clear account of these fields of physics.

The truth is, I hate Einstein and Hawking. I don't like them because I find it hard to use their mathematical ideas to deduce the theories I desire. It was that hard for me to classify where in the scientific community I fall, at first I thought that my ideas where into the quantum gravity field section but this was a lie. The quantum theory of gravity has not been fully settled. It was yesterday that I realized that my ideas fell into the Semi-Classical physical regime when I browsed it online;

"Semi-classical physics refers to a theory in which one part of a system is described quantum-mechanically whereas the other is treated classically" In general, it incorporates a development in powers of Planck's constant, resulting in the classical physics of powers 0, and the first nontrivial approximation into the powers of -1. (Wikipedia)

I am sorry, Semi-classical physics hasn't gained much interest, there are too many criticism about its meaning, researches into the field have been discouraged, few physicists have written about it and it is that unimportant. But anyway I am an amateur to venture into a field that is irrelevant. I don't give a damn what you think.

My first insight into the field of Semi- classical physics is traced back in 2010 in my first paper I published on arXiv.org titled "A hypothetical investigation into the realm of the microscopic and macroscopic universes beyond the standard model" This paper clearly shows that I was into the field without knowing. For sure I thought I was dealing with the field of Quantum Gravity by then.

Well, if you don't understand Semi-classical physics, Amateurs do. Below I show you why I think I understand the field and you surely do. I provide many ideas which I think the entire scientific community must investigate.

The meaning of semi-classical physics to an amateur

Assuming an experiment where the classical electric force f_e is balanced over the classical gravitational force f_g to determine their strength, the result will show that, the ratio of the two forces will follow a power law in powers of n of the gravitational coupling constant as,

$$\frac{f_e}{f_g} = \alpha_g^{\,n}$$

The left hand side of the equation represents the classical part of the system while the right hand side represents the quantum mechanical part of the system.

Let the classical part be described by two constants;

G- The Universal gravitational constant

c- The constant speed of light

Into (G, c)

Let the Quantum mechanical part be described by two constants,

$\hbar$- The reduced Planck constant

c-The constant speed of light

Into ($\hbar$, c)

Then from the above assumption Semi-classical physics will reduce results combining the constants above into (G, c, $\hbar$)

From the above formula we can deduce the time and length units of measure formulas to help us understand the field better,

$$\text{Time } t_n = \frac{Gm}{c^3} \alpha_g^{-n}$$

$$\text{Length } l_n = \frac{Gm}{c^2} \alpha_g^{-n}$$

Where m denotes the mass of a particle or body and $\alpha_g = \frac{Gm^2}{\hbar c}$ is the gravitational coupling constant. You can also assume interactions involving the electromagnetic coupling constant. The different fields of physics resulting from the above classification for different powers of n from 0, 1, 2 and -1/2 are given below,

For n=0

Classical General Relativity

$$t_0 = \frac{Gm}{c^3}$$

$$l_0 = \frac{Gm}{c^2}$$

For n=1

Quantum mechanics

$$t_1 = \frac{\hbar}{mc^2}$$

$$l_1 = \frac{\hbar}{mc}$$

For n=2

Semi-classical gravity

$$t_2 = \frac{\hbar^2}{Gcm^3}$$

$$l_2 = \frac{\hbar^2}{Gm^3}$$

For n= -1/2

Planck Units

$$t_{-1/2} = \left(\frac{G\hbar}{c^5}\right)^{1/2}$$

$$l_{-1/2} = \left(\frac{G\hbar}{c^3}\right)^{1/2}$$

The above derivation gives out a clear description of Semi-classical physics to a lay person.

One can decide to use n as a spatial dimension of space.

Applications of semi-classical physics
Radiation intensity of a black hole

<u>The classical part of a system</u>

Let the classical total force on an electron in orbit at a distance r from the nucleus of an atom be related to its electromagnetic and gravitational forces by,

$$f = \frac{F_G F_e}{F_B}$$

Where F_G is the gravitational force, F_e is the electric force and $F_B = Bev$ is the magnetic force

The angular momentum of an electron is given classically as,

$$L = \frac{Gm^2}{c} = mvr$$

<u>The Quantum mechanical part of the system</u>

The angular momentum is quantized as,

$$L = \frac{K_g e^2}{c} = \hbar$$

On eliminating the constant speed of light c from both the expression of the angular momentums we have

$$mvr = \frac{F_G}{F_e}\hbar$$

The ratio $\frac{F_G}{F_e}$ represents the classical part of the system while $\hbar$ represents the quantum part.

Eliminating F_G from the above expression we get the magnetic power as,

$$F_B c = \frac{2\pi r^2 \lambda mv F_g^2}{\hbar^2}$$

But the de Brogile wave length of an electron is $\lambda = h/mv$ and the surface area of the sphere of orbit of an electron is $A = 4\pi r^2$. Then the electromagnetic Intensity is given as,

$$I = \frac{F_B c}{A} = \frac{F_g^2}{2\hbar}$$

Thus the intensity of a wave is proportional to the square of the electric force If we let the power of the electromagnetic wave be P= F_{BC}, and n be the fine structure constant $\alpha = ke^2/\hbar c$, then the equation for the intensity of the classical electromagnetic wave comes out clearly as,

$$P = EB/\mu_0 = 2\varepsilon_0 E^2 c,$$

Where μo is the permeability of free space

Assuming the coupling of the forces to be,

$$\frac{f_e}{f_g} = \alpha_g{}^n$$

Then at n = -1, and $f_g = \frac{c^4}{8\pi G}$ we have the electric force as,

$$f_e = \frac{\hbar c^5}{8\pi G^2 m^2}$$

Then the intensity of the radiations will be given as

$$I = \frac{f_e^2}{2h} = \frac{\hbar c^{10}}{256\pi^3 G^4 m^4}$$

This expression comes from treating the particle classically in one part and then quantum mechanically in another part. It can be clearly seen above, that we haven't used the mathematics of general relativity or quantum field theory to reach at the result.

The earliest period of time in the history of the universe

<u>Classical part of the system</u>

Let the acceleration due to gravity of a particle (say an electron) in the gravitational field be given as

$$g = R/\Delta t^2$$

Where is Δt the time and R is the distance of the particle from the source. If the particle radiates energy then the energy per unit time is,

$$P = c^5/G$$

Quantum mechanical part of the system

The power and time must be quantized in units of $\hbar = h/2\pi$ where h is Planck constant, hence

$$P\Delta t^2 = n^2\hbar$$

Where n= 1,2,3…….. is the principle quantum number.

But the potential energy of the electron in the various energy states is,

$$W = -ke^2/R$$

where k is the Coulomb constant and e is the elementary charge. Since Δt² is known from the expression for acceleration due to gravity. Then the distance R is,

$$R = n^2 Gg\hbar/c^5$$

From which the total energy is given by,

$$W = -ke^2c^5/n^2Gg\hbar$$

From the Bohr-Einstein frequency (f) condition, applied to a transition from a level with $n = n_i$ to a level with $n = n_f$, The energy of a photon emitted by a hydrogen atom is given by the difference of two hydrogen energy levels

$$hf = E_i - E_f$$

Since frequency $f = c/\lambda$, where λ is the wavelength. Then we have,

$$1/\lambda = [ke^2c^4/2\pi G\hbar^2][1/g][1/n_f^2 - 1/n_i^2]$$

The equation obtained above shows some how a great significance of gravity in the quantum theory. So far it states that regardless of the levels in the transitions of an atom the acceleration due to gravity of the particles in the atom do greatly affect the nature of its spectrum.

The quantity $[ke^2c^4/ 2\pi G\hbar^2]$ in the formula above is the inverse of the square of time t and therefore,

$$1/t^2 = [ke^2c^4/ 2\pi G\hbar^2]$$

From which the time is obtained as $t = 1.58873 \times 10^{-42}$s. This is the earliest period of time in the history of the universe.

The Weidmann Franz- Lorenz law

Treating one part of the system classically (macroscopic) and the other quantum mechanically (microscopic), we have the formula for the electric force acting on an electron in motion as

$$F = \frac{n^2}{\alpha_g} f_g$$

Where n, is the principle quantum number.

The above formula differs from the one previously given. On squaring the above equation we obtain the square of the electric field as,

$$E^2 = \frac{n^4 c^4}{G^2 s^2 m^2} \left(\frac{c^3 \hbar}{8\pi Gm}\right)^2$$

From the formula for the temperature of the black hole, the function $\frac{c^3 \hbar}{8\pi Gm}$ is related to temperature as kT, and then the law for thermal conductivity will be reduced as,

$$\frac{\pi^2 E^2 G^2 m^2}{3 T c^4} = \left(\frac{n^4 \pi^2}{3}\right) \left(\frac{k}{s}\right)^2 T$$

The left hand side represents the ratio of the thermal conductivity K to the electric conductivity δ. The right hand side is the Weidman –Franz law. Therefore the left side of the equation represents the macroscopic part of the system while the right hand side represents the microscopic part of the system. Then the left-hand side will be given as,

$$\frac{K}{\delta} = \frac{1}{3}\left(\frac{\pi Gm}{c^2}\right)^2 \frac{E^2}{T} = \frac{\pi A\, E^2}{3\, T}$$

Where A is the surface area of a body $A = \pi r_s^2$ with the schwarzichild's radius r_s. This is the conductivity ratio of a black hole.

The Art of Reductionism

Scientific reductionism is the idea of reducing complex interactions and entities to the sum of their constituent parts, in order to make them easier to study (explorer.com). It is based on the idea that science can be used to explain everything by a mere look at the individual constituent processes.

There are three types of reductionism, that is, ontological, methodological and theory reduction. In this section we shall emphasize theory reduction because we have a great deal of reducing known laws of physics from a somewhat simple rule. This was the case when Kepler's laws of the motion of planets and Galileo's theories of motion were reduced to the Newtonian theories of mechanics.

Newtonian Mechanics became a more general theory simply because all the explanatory power of Kepler's and Galileo's laws was contained in it. Therefore theoretical reduction is considered as the reduction of one explanation or theory to another.

The most interesting thing about this section is that, during the process of reduction we create a relationship between the known law to another law explaining the same thing but unknown to the entire physics community. For example in the reduction of the Weidman Franz- Lorenz law we create in a process a law for the thermal conductivity of gravito-electric phenomenon.

Therefore reductionism is deriving something complicated from something simple. For example in the derivation of the Weidman Franz law we set a formula that states that, the electric force (Ee) on an electron is proportional to the gravitational force at the schwarzichild's radius $\left(\frac{c^4}{8\pi G}\right)$ but inversely proportional to the gravitational coupling constant $\left(\frac{Gm^2}{\hbar c}\right)$ as given below,

$$F = \frac{n^2}{\alpha_g} f_g$$

Where n, is the principle quantum number.

On squaring the above equation we obtain the square of the electric field as,

$$E^2 = \frac{n^4 c^4}{G^2 e^2 m^2} \left(\frac{c^3 \hbar}{8\pi G m}\right)^2$$

From the formular for the temperature of the black hole, the function $\frac{c^3 \hbar}{8\pi G m}$ is related to temperature as kT, and then the law for thermal conductivity will be reduced as,

$$\frac{\pi^2 E^2 G^2 m^2}{3 T c^4} = \left(\frac{n^4 \pi^2}{3}\right) \left(\frac{k}{e}\right)^2 T$$

The left hand side represents the ratio of the thermal conductivity K to the electric conductivity δ, which is the Weidman Franz law. From the above reduction we have generated an important rule given by,

$$\frac{K}{\delta} = \frac{1}{3}\left(\frac{\pi Gm}{c^2}\right)^2 \frac{E^2}{T} = \frac{\pi^2 r_s^2}{3}\frac{E^2}{T}$$

The above formula explains the thermal properties of black holes at the schwarzichild's radius r_s.

onstruction of a Consistent Physical Theory of Nature

A consistent theory of nature, simply the "theory of everything" is constructed using one of the profound ideas of "axioms", that, when the Stoney units of measure are multiplied by the coupling constant (a dimensionless number) of a form $\alpha^{\frac{n-1}{2}}$, one can easily calculate the mass of all particles in the universe and their length of time scales with accuracy provided the value of n is known. The mass of the electron is calculated at $\alpha=1/137.036$ and $n=21.32$ while the mass of the earth is calculated at $n= -29.99$, hence solving one of the major unsolved problems in physics. The Planck mass is calculated and determined in principle to be 5.4556×10^{-8} Kg, a different value from the given value would lead to variations in our fundamental physical constants of electricity and gravity. The energy scales at given length scales in literature are also deduced in which a requirement to revisit our profound known physical theories is proposed.

One of the major unsolved problems in physics is developing a final theory, ultimate theory or theory of everything. In this paper we present a series of hypotheses and speculations leading inescapably to a conclusion that when the Stoney fundamental units of measure are multiplied by the electromagnetic coupling constant (fine structure constant) powered by any integer, $a^{\frac{n-1}{2}}$ from 0,1,2,..................,n, one gets to calculate the mass of all particles in the universe, the lengths between them and the time expressible at a scale of the known fundamental physical constants of nature. Our hypotheses may be wrong and our speculations idle, but the uniqueness and simplicity of our scheme are reasons enough that it be taken seriously.

Our starting point is the assumption that all of the fundamental physical units of measure can be calculated and organized to demonstrate different branches and scales of physics whatsoever using the following formulas,

Length

$$L_n = \frac{e}{c^2}\sqrt{\frac{G}{2\varepsilon_0}}\,a^{n-1} \qquad (140)$$

Time

$$t_n = \frac{e}{c^3}\sqrt{\frac{G}{2\varepsilon_0}}\,a^{n-1} \qquad (141)$$

Mass

$$M_n = e\sqrt{\frac{1}{2G\varepsilon_0}}\,a^{n-1} \qquad (142)$$

Where α is the coupling constant for either electromagnetic or gravitational interactions, G is the universal gravitational constant, e is the elementary charge on an electron, c is the speed of light and ε_0 is the permittivity of free space, the meaning of n is left to be investigated as per the meaning of the theory.

case1:

We derive the fundamental units of measure at values of n=0, 1,2,3,4 and 5 only for the fine structure constant $\alpha = \frac{e^2}{4\pi\varepsilon_0 \hbar c}$ where is the reduced Planck constant $\hbar = \frac{h}{2\pi}$.

<u>At n=0</u>

$$L_o = \sqrt{\frac{2\pi G \hbar}{c^3}} \quad , \quad t_o = \sqrt{\frac{2\pi G \hbar}{c^5}} \quad , \quad M_o = \sqrt{\frac{2\pi \hbar c}{G}}$$

<u>At n=1</u>

$$L_1 = \frac{e}{c^2}\sqrt{\frac{G}{2\varepsilon_0}} \quad , \quad t_1 = \frac{e}{c^3}\sqrt{\frac{G}{2\varepsilon_0}} \quad , \quad M_1 = \frac{e}{\sqrt{2G\varepsilon_0}}$$

At n=2

$$L_2 = \frac{e^2}{\varepsilon_0}\sqrt{\frac{G}{8\pi c^5 \hbar}} \quad , \quad t_2 = \frac{e^2}{\varepsilon_0}\sqrt{\frac{G}{8\pi c^7 \hbar}} \quad , \quad M_2 = \frac{e^2}{\varepsilon_0}\sqrt{\frac{1}{8\pi G \hbar c}}$$

At n=3

$$L_3 = \frac{e^3}{\pi c^3}\sqrt{\frac{G}{32\varepsilon_0^3}} \quad , \quad t_3 = \frac{e^3}{\pi c^4}\sqrt{\frac{G}{32\varepsilon_0^3}} \quad , \quad M_3 = \frac{e^3}{\pi \hbar c}\sqrt{\frac{1}{32\varepsilon_0^3 G}}$$

At n=4

$$L_4 = \frac{e^4}{\varepsilon_0^2}\sqrt{\frac{G}{128\pi^3 \hbar^3 c^7}} \quad , \quad t_3 = \frac{e^4}{\varepsilon_0^2}\sqrt{\frac{G}{128\pi^3 \hbar^3 c^9}} \quad ,$$

$$M_3 = \frac{e^4}{\varepsilon_0^2}\sqrt{\frac{1}{128\pi^3 G \hbar c}}$$

At n=5

$$L_S = \frac{e^5}{\pi^2 \hbar^2 c^4} \sqrt{\frac{G}{512\varepsilon_0^5}} \quad , \quad t_S = \frac{e^5}{\pi^2 \hbar^2 c^5} \sqrt{\frac{G}{512\varepsilon_0^5}} \quad ,$$

$$M_3 = \frac{e^5}{\pi^2 \hbar^2 c^2} \sqrt{\frac{1}{512 G \varepsilon_0^5}}$$

At n=0, we obtain the Planck natural units while at n=1, we obtain the Stoney units of measure

case2:

We further derive the fundamental units of measure at values of n=0, 1, 2, only for the gravitational coupling
$$\alpha = \frac{Gm^2}{\hbar c}$$

<u>At n=0</u>

$$L_o = \frac{e}{m}\sqrt{\frac{\hbar}{2\varepsilon_0 c^3}} \quad , \quad t_o = \frac{e}{m}\sqrt{\frac{\hbar}{2\varepsilon_0 c^5}} \quad , \quad M_o = \frac{e}{Gm}\sqrt{\frac{\hbar c}{2\varepsilon_0}}$$

<u>At n=1</u>

$$L_1 = \frac{e}{c^2}\sqrt{\frac{G}{2\varepsilon_0}} \quad , \quad t_1 = \frac{e}{c^3}\sqrt{\frac{G}{2\varepsilon_0}} \quad , \quad M_1 = \frac{e}{\sqrt{2G\varepsilon_0}}$$

At n=2

$$L_2 = meG\sqrt{\frac{1}{2\varepsilon_0 c^5 \hbar}} \quad , \quad t_2 = meG\sqrt{\frac{1}{2\varepsilon_0 c^7 \hbar}} \quad ,$$

$$M_2 = em\sqrt{\frac{1}{2\varepsilon_0 \hbar c}}$$

It proves difficult to deduce the Planck units here, simply because the charge and mass do not cancel out. But if you set the ratio of charge to mass at n=0 in the above formulas as $\frac{e}{m} = \sqrt{4\pi\varepsilon_0 G}$, one obtains the Planck units. Also, one obtains the values of n=2 in Case1 when we substitute for $m = \frac{e}{\sqrt{4\pi\varepsilon_0 G}}$, in Case2, for n=2. This means that, the formulas which do not exist in case2 but are present in case1 can be calculated by applying a simple formula, $e = m\sqrt{4\pi\varepsilon_0 G}$ and vise versa is true.

When we substitute for $e = m\sqrt{4\pi\varepsilon_0 G}$, at n=1, we obtain,

$$L_1 = \frac{Gm}{c^2}\sqrt{2\pi}, \quad t_1 = \frac{Gm}{c^3}\sqrt{2\pi}, \quad M_1 = m\sqrt{2\pi}$$

This represents formulae at a scale of general relativity, in which it is deduced here that, the mass of a particle in both the special and general relativity theory makes sense when multiplied by a constant $\sqrt{2\pi}$.

When $me = m^2\sqrt{4\pi\varepsilon_0 G}$ at n=2 above, we obtain

$$L_2 = m^2\sqrt{\frac{2\pi G^s}{c^5\hbar}}, t_2 = m^2\sqrt{\frac{2\pi G^s}{c^7\hbar}}, M_2 = m^2\sqrt{\frac{2\pi G}{\hbar c}} = \frac{2\pi m^2}{m_p}$$

Where m_p is the Planck mass $\sqrt{\frac{2\pi\hbar c}{G}}$

It should however be taken seriously from the above investigation that changing the number 2π in the formulas (case1, at n=0, the planck units/scales), will change the statement of the formula $e = m\sqrt{4\pi\varepsilon_0 G}$, which will mean that the values of the fundamental physical constants $\frac{1}{4\pi\varepsilon_0}$, G are varying, therefore in order to maintain the constants unchanged we have to maintain the Planck units unchanged in formula as they are derived here. Thus the Planck mass will have a mass given by, $2.2176470119 \times 10^{-8} \sqrt{2\pi} = 5.4556 \times 10^{-8}$ Kg.

At present there is no candidate theory of everything that includes the standard model of particle physics and general relativity. For example, no candidate theory is able to calculate the mass of an electron. However in this paper the mass of an electron is deduced when n=21.32 and α=1/137.036 as,

$$M_{21.32} = e\sqrt{\frac{1}{2G\varepsilon_0}\left(\frac{1}{137.036}\right)^{20.32}} = 9.082073363 \times 10^{-31} kg$$

$$L_{21.32} = 6.745125 \times 10^{-58} m$$

$$t_{21.32} = 2.25 \times 10^{-66} s$$

Also the proton mass is deduced at n= 18.26, and α=1/137.036 as,

$$M_{18.26} = e\sqrt{\frac{1}{2G\varepsilon_0}\left(\frac{1}{137.036}\right)^{17.26}} = 1.688659377 \times 10^{-27} kg$$

Other masses including the mass of the earth (n= -29.99) can be deduced in the same way. It is important to note that, the value of n is negative for massive particles (e.g mass of the Sun and earth) but positive for microscopic particles like electrons and protons.

It is hereby noted that the values of the energy scales corresponding to the given length scale are off the scale and do not necessarily represent phenomenon at each given length scale. The values of these energy scales for each interaction as quoted in scientific literature will prove to be different from the ones represented here. For example;

For atomic length scale with $l_a \sim 10^{-10} m$, the value of n to be used in calculating other scales will be given by n= -22.83, from which the energy scale can be calculated as,

$$E_{-22.83} = M_{-22.83} c^2 \sim 7.6 \times 10^{43} GeV$$

For strong interaction length scale with $l_s \sim 10^{-15} m$, the value of n to be used in calculating other scales will be given by n= -18.151, from which the energy scale can be calculated as,

$$E_{-18.151} \sim 7.6 \times 10^{38} GeV$$

For electroweak interaction length scale with $l_w \sim 10^{-18} m$, the value of n to be used in calculating other scales will be given by n= -15.343, from which the energy scale can be calculated as,

$$E_{-15.343} \sim 7.6 \times 10^{35} \, GeV$$

It is possible that the correct theory of everything has been found in the formulas given above. It therefore seems appropriate for the reader or researchers to consider the calculation and determination of the values of the masses of all particles at given length/time scales in the universe with explicit accuracy, even if some people may consider such an enterprise premature or foolhardy. It is worth noting that the length and time scale through which one can probe the whole mass of the earth is $4.439 \times 10^{-3} m$ and $1.4806 \times 10^{-11} s$ respectively. It is therefore important to know how one can calculate the mass of any particle with accuracy and then inquire with simplicity into the length and time scale at which such a particle can be studied. This then means that any consistent theory of nature like the one constructed would be able to deduce the required derived quantities (i.e. voltage, current, magnetic field etc) from the given formulas for fundamental physical units of measure of mass, length and time without a need to inquire into other theories like the standard model, string theory or quantum gravity.

Is It Possible That There Is A Universe In Every Particle?

By definition an atom is the smallest unit of an element that retains the chemical properties of that element. An atom has an electron cloud consisting of negatively charged electrons surrounding a dense nucleus.

For so long scientists have been extracting electrons from atoms and bringing them onto the earth. Remember, atoms have been into existence onto the earth for many years. To me they seem like small universes within a big universe and a combination of them forms the universe within which we live. There can be ways for us to travel to those atoms and see the life there. You think this is fiction, no you are wrong. Look if the atoms merge to form the universe then the universe will separate into small parts to form the atom and this will reduce our size and shape such that we have the dimensions of the universe to which we can survive as you see it today.

If these electrons have the ability to live in the universe then also there must be a possibility for man to live on an atom. For example consider a person going to stay on an atom, remember the atom is very small and man can't even stand on it that is what you think. You will also think that he can't even see where he will land because he can't see the atom with his naked eyes. But the solution is here for a person to stay on the atom.

(i) The first ideological formulation of the possibility is by assuming that we are small towards an atom and an electron is big towards the planet.

(ii) The second assumption is that, gravity exists on an atom

(iii) The third assumption is that the space surrounding the atom is in a frozen state , that is solid like and man can move on it without floating into space

(iv) The fourth is that the life on the atom for both plant and man is constant

From the above assumptions, you can now imagine bringing a small particle like an electron from the atom onto the earth. Of course it would be easier when you take the above assumptions to be true.

(i) The electron it's self will assume to be big for it to stay on the earth and so it does (ii)It will also assume that the space is in a frozen state for it to move on it

(ii)It will also assume that life is constant and extra

The above assumptions are true only if you set yourself into motion to the atom. The atom may be an interesting place for you than our universe. **But where is the way to it?**

The way to the atom is your imagination or simply what you think. See if you imagine that you are a small particle then that what you are but when you think that you are a big particle then the impression is true that you are. To go to the atom takes a few seconds than going to the moon. There is of course no transport but distraction of the body.

Your body will break into small particles where by each will be able to land on the atom and finally they will merge at exactly the same dimension as the atom that will make your size and shape fill comfortable on it.

Newton's Biggest Blunder: Re-defining Gravity

It is known that Newton deduced his law of universal gravitation from Keplers laws. What if Kepler made a mistake in his calculations, What if the same mistake is still hidden in Newton's law? What if general relativity failed to determine the same error and instead brought complications? What is the error, what is this dimensionless constant (a number)? Does uncovering this error lead to a complete quantum theory of gravity? Is this error or number the missing link between gravity and quantum mechanics?

Below I re-define gravity in which I rediscover the missing number, a dimensionless physical quantity in Newton's law of gravitation.

Rule 1

The ratio of the velocity of a particle in motion (v) to the speed of light (c) must be eqaul to the ratio of the potential energy of the particle (W_o) to the energy carried by the particle (photon) of light (W)

$$\frac{v}{c} = \frac{W_o}{W}$$

Orbits around the Sun

From rule one a particle will move with a velocity relative to that of light. The radiations or electromagnetic waves emitted by the sun contain changing electric E and magnetic fields B related to the speed of the electromagnetic waves (light) and then rule one becomes

$$\frac{Bv}{E} = \frac{W_o}{W}$$

Since the energy carried by the photon of light is related to the period of revolution of the particle in orbit $\frac{h}{T}$, where h is the Planck constant and the potential energy of the particle in orbit around the sun in the gravitational field of the sun is also related to its mass m and the distance r between the sun and the particle, then the period of revolution of the particle around the sun is

$$T = \frac{Bvhr}{EGM^2}$$

Where G- Universal gravitational constant

It must also be true that the electric field experienced by the particle in orbit around the sun due to the sun's electric force is proportional to the charge q on the particle in orbit but falls with the square of the distance from the sun, and then the period of orbit is

$$T = \frac{4\pi\varepsilon_0 Bvhr^3}{qGM^2}$$

Where, ε_0 - the permittivity of free space

Since particles exhibit wave properties then the velocity of the particle in orbit is related to its wavelength and period of orbit by $v = \frac{\lambda}{T}$, from which the we get the square of the period as

$$T^2 = \frac{4\pi^2}{GM} \left(\frac{2\varepsilon_0 B \lambda \hbar}{Mq}\right) r^3$$

Where $\hbar = \frac{h}{2\pi}$

This is the advanced or modified Kepler's periods (Harmonic Law). The quantity in the brackets $\left(\frac{\varepsilon_0 B \lambda \hbar}{Mq}\right)$ is a dimensionless constant and it is what is missing in Newton's derivation of Kepler's Laws.

Of The Galactic And Atomic

When the dimensionless constant is unity, we have the galactic and atomic descriptions of reality separated. In other words we have to parts of the universe that are separated and need to be merged. The modified Kepler harmonic law given above is a true description of the universe involving all the parts that is, the small and large particles. The galactic is when the dimensionless constant is unity. The atomic is given when

$$\left(\frac{2\varepsilon_0 B\lambda\hbar}{Mq}\right) = 1$$

The wavelength of which is

$$\lambda = \frac{Mq}{2\varepsilon_0 B\hbar}$$

For $B = \frac{E}{v}$ and $E = \frac{q}{4\pi\varepsilon_0 r^2}$ we have

$$\frac{4\pi^2 r^2}{\lambda} = \frac{h}{Mv}$$

Rule 2

The circumference of the smallest allowed orbit for the particle is equal to the wavelength of the particle. This statement is a proof for the quantized energy of the particle in an atom and is expressed as $2\pi r = \lambda$. With this rule, the wavelength of the particle is therefore related to its momentum mv by

$$\lambda = \frac{h}{Mv}$$

It is therefore true according to deBrogile and the derivation given here that particles also have wave properties.

From the derivations given here it is clear that we have a description for both the microscopic and macroscopic particles.

The modified Newton's universal law of gravitation (MONG)

Following the same procedure that Newton used in the derivation of Kepler's harmonic law, the advanced law derived here can be deduced using the advanced Gravitational force given as

$$MONG\ FORCE = \frac{GM^2}{r^2}\left(\frac{Mq}{2\varepsilon_0 B\lambda\hbar}\right)$$

This is arranged as

$$\text{MONG FORCE} = \frac{4\pi^2 G k_e q M^3}{r^2}\left(\frac{1}{B\lambda h}\right)$$

Where $k_e = \frac{1}{4\pi\varepsilon_0}$ - the coulomb constant

For $2\pi r = \lambda$

$$\text{MONG FORCE} = \frac{2\pi G k_e q M^3}{r^3}\left(\frac{1}{Bh}\right)$$

The application of the above force is given (see Equation k) where it is used to derive the equation for the entropy of the black hole.

Space-time Singularity or Quantum Black Holes?

It has been known for some time that a star more than three times the size of our Sun collapses in this way, the gravitational forces of the entire mass of a star overcomes the electromagnetic forces of individual atoms and so collapse inwards. If a star is massive enough it will continue to collapse creating a Black hole, where the whopping of space time is so great that nothing can escape not even light, it gets smaller and smaller. The star in fact gets denser as atoms even subatomic particles literally get crashed into smaller and smaller space, and its ending point is of course a space time singularity.

In summary, a Black hole is that object created when a dying star collapses to a singular point, concealed by an event horizon, it is so dense and has strong gravity that nothing, including light, can escape it. Black holes are predicted by general relativity, and though they cannot be "seen," several have been inferred from astronomical observations of binary stars and massive collapsed stars at the centers of galaxies.

Black holes formed by gravitational collapse require great energy density but there exists a new breed of Black holes that where formed in the early universe after the big bang, where the energy density was much greater allowing the formation of Primordial Black holes with masses ranging from, $10^8, 10^{12} - 10^{17} kg$. Therefore the formation of primordial, min or quantum black holes was due to density perturbations forming in it a gravitational collapse in the early universe.

A Black hole might not actually be a physical object in space but rather a mathematical singularity, a prediction of Einstein's General Relativity theory, a place where the solutions of Einstein differential equations break down. A space-time singularity therefore is a position in space where quantities used to determine the gravitational field become infinite; such quantities include the curvature of space-time and the density of matter. Singularities are places where both the curvature and the energy-density of matter become infinitely large such that light cannot escape them. This happens for example inside black holes and at the beginning of the early universe.

Singularities in any physical theory indicate that either something is wrong or we need to reformulate the theory itself. Singularities are like dividing something by zero. The problems in General relativity arise from trying to deal with a point in space or a universe that is zero in size (infinite densities). However, quantum mechanics suggests that there may be no such thing in nature as a point in space-time, implying that space-time is always smeared out, occupying some minimum region. The minimum smeared-out volume of space-time is a profound property in any quantized theory of gravity and such an outcome lies in a widespread expectation that singularities will be resolved in a quantum theory of gravity. This implies that the study of singularities acts as a testing ground for quantum gravity.

Loop quantum gravity (LQG) suggests that singularities may not exist. LQG states that due to quantum gravity effects, there must be a minimum distance beyond which the force of gravity no longer continues to increase as the distance between the masses become shorter or alternatively that interpenetrating particle waves mask gravitational effects that would be felt at a distance. It must also be true that under the assumption of a corrected dynamical equation of LQ cosmology and brane world model, for the gravitational collapse of a perfect fluid sphere in the commoving frame, the sphere does not collapse to a singularity but instead pulsates between a maximum and minimum size, avoiding the singularity.

Additionally, the information loss paradox is also a hot topic of theoretical modeling right now because it suggests that either our theory of quantum physics or our model of black holes is flawed or at least incomplete. and perhaps most importantly, it is also recognized with some prescience that resolving the information paradox will hold the key to a holistic description of quantum gravity, and therefore be a major advance towards a unified field theory of physics.

Singularities are a sign that the theory breaks down and has to be replaced by a more fundamental theory. And we think the same has to be the case in General Relativity, where the more fundamental theory to replace it is quantum gravity.

If black holes are as a result of the solutions to the Einstein's differential equations breaking down, then what is real?

Whether in gravitational collapse or the early universe, we now know that the formation of Black holes or space time singularities requires great and much greater energy density. This we know because while the left hand side of Einstein field equations representsnts the metric of space-time curvature, the right hand side represents the matter- energy content of the classical matter fields of pressure and energy density. This therefore means that quantum mechanics which plays an important role in the behavior of the matter fields has no place in the Einstein field equations and this is what brings on the singularities that plague the general relativity theory.

$$G_{\mu v} + \Lambda g_{\mu v} = \frac{8\pi G}{c^4} T_{\mu v}$$

Because of this, one therefore has a problem of defining a consistent scheme in which the space time metric is treated classically but is coupled to the matter fields which are treated quantum mechanically.

What is not real is to use the stress energy tensor (classical pressure and energy density) on Black holes instead of the quantum mechanical energy density.

The approximation I shall use on my journey to quantum gravity (Quantum Black holes) is that the matter fields, such as scalar, electro-magnetic, or neutrino fields, obey the usual wave equations with the left hand side replaced by a classical space time second order curvature ($\Lambda = \frac{1}{R^2}$), where R is the radius of curvature) while the right hand stress-energy tensor replaced by the quantum mechanical energy density ($\rho = \frac{F^2}{8\pi\alpha\hbar c}$ (1)) Where F is the force involved in an interaction α is the coupling constant that determines the strength of the force, and ℏ is the reduced Planck constant. The equation represents the coupling constant (α) as a function of the energy density (ρ) for any force (F) exerted in an interaction. The application of this equation is the Franzl Aus Tirol curve on Wikipedia's "Coupling constant". Another application is the derivation of energy stored in the electromagnetic field (see Appendix 1).Therefore the general theory of quantum mechanics in curved space –time will be given by this simple equation, $\Lambda = \frac{8\pi G}{c^4}\rho$, where

$$\Lambda = \frac{GF^2}{\alpha\hbar c^5} = \frac{F^2}{\alpha E_{pl}^2} \qquad (2)$$

Where, $E_{pl} = M_{pl}c^2$ is the Planck energy and M_{pl} is the Planck mass

From the above given equation we see that high space curvature will always be achieved when the square of the force involved increases. According to the theory given, this will only occur at the Planck energy level where space is discrete or granular in nature (its building blocks being exactly the Planck mass, simply put, the atoms of space). There is no change in energy because the only energy involved in the process is the constant Planck energy of the Planck mass.

As we said earlier, that the formation of a black hole due to the process of gravitational collapse occurs in the presence of great energy density and also that the formation of primordial black holes in the early universe occurs in the presence of a much greater energy density, our theory suggests that this energy density is high because of the strong gravitational force involved in the process. According to general relativity, this force is a constant and is given by, $F = \frac{c^4}{G}$. Therefore from equation (2), when this force is present the curvature of space scales as the inverse of the square of the Planck length,

$$\Lambda = \frac{c^3}{\alpha\hbar G} = \frac{1}{\alpha l_p{}^2} \qquad (3)$$

Where $l_p = \sqrt{\frac{\hbar G}{c^3}}$ is the Planck length.

This implies that, in the theory of quantum mechanics in curved space-time for the gravitational collapse of a star, the star does not collapse to a singularity but instead to a Planck sized star of Planck length close to $10^{-35}m$ and this will happen only when $\alpha = 1$. Finally, in the theory of quantum mechanics in curved space-time, we consider the possibility that the energy of a collapsing star and any additional energy falling into the hole could condense into a highly compressed core with density of the order of the Planck density. Since the energy density or pressure is expressed as in equation (1),

$$\rho = \frac{F^2}{8\pi\alpha\hbar c}$$

Therefore nature appears to enter the quantum gravity regime when the energy density of matter reaches the Planck scale. The point is that this may happen well before relevant lengths become planckian. For instance, a collapsing spatially compact universe bounces back into an expanding one. The bounce is due to a quantum-gravitational repulsion which originates from the modified Heisenberg uncertainty, and is akin to the force that keeps an electron from falling into the nucleus. And from the uncertainity principle, this repulsion force is given by,

$$F = \frac{c^4}{G}$$

Therefore the bounce does not happen when the universe is of planckian size, as before; it happens when the matter energy density reaches the Planck density in this way,

$$\rho = \frac{c^7}{8\pi\alpha\hbar G^2} \qquad (4)$$

At this energy density, a Planck star is formed. The key feature of this theoretical object is that this repulsion arises from the energy density, not the Planck length, and starts taking effect far earlier than might be expected. This repulsive 'force' is strong enough to stop the collapse of the star well before a singularity is formed, and indeed, well before the Planck scale for distance. Since a Planck star is calculated to be considerably larger than the Planck scale for distance, this means there is adequate room for all the information captured inside of a black hole to be encoded in the star, thus avoiding information loss.

The analogy between quantum gravitational effects on

Cosmological and black-hole singularities has been exploited to study if and how quantum gravity could also resolve the r = 0 singularity at the center of a collapsed star, and there are good indications that it does. For example, if we extend (3) to n extra dimensions we have,

$$R = \alpha^{n/2} l_p$$

Where α in this case is the size of the extra dimensions and α^n is the flux in the extra dimesions. Let the size of the extra dimension be given as the gravitational coupling constant, $\alpha = \dfrac{GM^2}{\hbar c} = \left(\dfrac{M}{M_{pl}}\right)^2$, then the size of a star will be given by,

$$r = \left(\frac{M}{M_{pl}}\right)^n l_p \qquad (5)$$

Where M is the mass of the star and n is positive. For instance, if n = 1/3, a stellar-mass black hole would collapse to a Planck star with a size of the order of 10^{-10} centimeters. This is very small compared to the original star in fact, smaller than the atomic scale but it is still more than 30 orders of magnitude larger than the Planck length. This is the scale on which we are focusing here. The main hypothesis here is that a star so compressed would not satisfy the classical Einstein equations anymore, even if huge compared to the Planck scale. Because its energy density is already planckian.

What is real is that; the gravitational collapse of a star does not lead to a singularity but to one additional phase in the life of a star: a quantum gravitational phase where the gravitational attraction is balanced by a quantum pressure.

What is real? Is it Volume or Area Entropy Law of Black Holes?

The development of general relativity followed a publication of acceleration under special relativity in 1907 by Albert Einstein. In his article, he argued that any mass will "Distort" the region of space around it so that all freely moving objects will follow the same curved paths curving toward the mass producing the distortions. Then in 1916, Schwarzschild found a solution to the Einstein field equations, laying the groundwork for the description of gravitational collapse and eventually black holes.

A black hole is created when a dying star collapses to a singular point, concealed by an "event horizon;" the black hole is so dense and has such strong gravity that nothing, including light, can escape it; black holes are predicted by general relativity, and though they cannot be "seen," several have been inferred from astronomical observations of binary stars and massive collapsed stars at the centers of galaxies.

These objects have puzzled the minds of great thinkers for many years. History puts it that, they were first predicated by John Michell and Pierre-Simon Laplace in the 18th century but David Finkelstein was the first person to publish a promising explanation of them in 1958 based on Karl Schwarz child's formulations of a solution to general relativity that characterized black holes in 1916.

In 1971, Hawking developed what is known as the second law of black hole mechanics in which the total area of the event horizons of any collection of classical black holes can never decrease, even if they collide and merge. This was similar to the second law of thermodynamics which states that, the entropy of a system can never decrease.

Then in 1972 Bekenstein proposed an analogy between black hole physics and thermodynamics in which he derived a relation between the entropy of black hole entropy and the area of its event horizon.

$$S = \frac{Akc^3}{4G\hbar}$$

In 1974, Hawking predicted an entirely astonishing phenomenon about black holes, in which he ascertained with accuracy that black holes do radiate or emit particles in a perfect black body spectrum.

$$T = \frac{\hbar c^3}{8\pi GMk}$$

The Bekenstein-Hawking area entropy law raises a number of questions. Why does the entropy of a Black hole scale with its area and not with its volume? For systems that we have studied, the entropy is proportional to the volume of the system. If entropy is proportional to area, so what do we make of all those thermodynamic relations that include volume, like Boyle's law or descriptions for a gas in a box? In otherwords how do we associate volume to the entropy of a Black hole?

In the previous section we defined a Black hole as a mathematical spacetime singularity that is; a position in space where quantities used to determine the gravitational field become infinite; such quantities include the curvature of spacetime and the density of matter. That is, for high or infinite densites where matter is enclosed in a very small volume of space General relativity breaks down. Quantum mechanics suggests that there may be no such thing in nature as a point in space-time, implying that space-time is always smeared out, occupying some minimum region. The minimum smeared-out volume of space-time is a profound property in any quantized theory of gravity and such an outcome lies in a widespread expectation that singularities will be resolved in a quantum theory of gravity. Therefore associating area to the entropy of a black hole means that the black hole has no volume which may not be true in a theory of quantum mechanics in curved space-time.

If density is the amount of energy contained within a given volume of space, then a Black hole must have a density and its volume will be determined by the amount os space enclosed by the surface area of its event horizon. Then from our famous equation (1) we can derive the volume entropy law of a Black hole. Let the energy density of a black hole be given as the work done on the system by exterior agents (W) per unit volume (V),

$$\rho = \frac{W}{V} = \frac{F^2}{8\pi\alpha\hbar c}$$

Since the gravitational force of a black hole is very strong and is given by, $F = \frac{c^4}{G}$. Then the work done by exterior agents on a black hole is related to its volume as,

$$W = \frac{c^7}{8\pi\alpha\hbar G^2} V = PV \qquad (6)$$

Where, P denotes the Planck pressure (or energy density) of a Black hole.

Since entropy is the quantity of heat or workdone (W) per unit temperature of a black hole (T) then from the following simple rule we derive the temperature of a Black hole as,

$$E_{pl} = NkT$$

Where, E_{pl} is the Planck energy, k is the Boltzmann constant and N is the Avagadro number given as,

$$N = \frac{W}{E_{pl}}$$

From which the temperature of a Black hole is given by

$$T = \frac{\alpha \hbar^2 G}{kc^2 V}$$

If the Black hole is a sphere with a Planck length $l_p = 1.616 \times 10^{-35} m$, then the volume of this Black hole is approximately, l_p^3. Therefore a Black hole with this volume will have the temperature $T = 1.416 \times 10^{32} K$ and this is exactly the earliest temperature in the history of the universe. Then finally the entropy of a Black hole will be given as

$$S = \frac{W}{T}$$

$$S = \frac{V^2 k c^9}{8\pi\alpha^2 \hbar^3 G^3} \approx \frac{V^2 k}{8 l_p^6} \qquad (7)$$

This reduces to the Bekenstein-Hawking area entropy law when the volume of the black hole is,

$$V = l_p^2 \sqrt{2A} \qquad (8)$$

As we said earlier the volume of a Black hole is determined by the amount of space enclosed by the surface area of its event horizon.

This section has presented a new approach to the entropy of Black holes. The major result of the research is the derivation of the volume entropy law that is different from the Bekenstein-Hawking area entropy law. As far as this book is concerned there is no other theory from which such a calculation can proceed. Hence the methods used in here are the only one from which a detailed quantum theory of gravity precedes and where the result of the volume entropy law can be achieved.

Is it Dark Matter, MOND or Quantum Black Holes?

Most physicists think that dark matter is a particle, similarly to the most subatomic particles that we know of like the protons, neutrons and electrons. Whatever it is, it behaves very similarly to gravity. But it doesn't emit or absorb light and it passes through normal matter undetected. We would like to know what particle it is? For example, how heavy is it? Or does anything at all happen if it interacts with normal matter? The only way we can answer the first two questions is through calculating from first principles the values of the cosmological constant as given by Planck collaborations (2018) and once our hypothesis matches with the Planck experiment then we will surely have proved what really dark matter and MOND is.

Consider the fabric of space-time to be made up of Planck mass relics or Primordial Black holes formed in the early universe after the big bang. Let these Planck particles which are the building blocks of space-time have a mass equal to the Planck mass of $M_{pl} = \sqrt{\frac{hc}{G}} = 2.18 \times 10^{-8} kg$.

Just like a gas in the box, these Planck particles are enclosed in the vaccum of space together with their antiparticle of negative mass $M_D = 2.76 \times 10^{-9} kg = 0.127 M_{pl}$.

The Planck particles are seen moving away from their antiparticle due to the Newtonian force law of $F = M_{pl} a_o$, where a_o is the low gravitational acceleration in MOND.

This movement of the Planck particles from their antiparticle is what we observe as an expanding space and hence the vacuum energy density or the cosmological constant from equation (1),

$$\rho = \frac{M_{pl}^2 a_o^2}{8\pi \alpha \hbar c} \qquad (9)$$

Since the gravitational coupling constant is known to be $\alpha = \frac{G M_D^2}{\hbar c}$, where M_D has a negative mass value. Putting this into account, we get the vacuum energy density value in agreement with the Planck collaboration 2018 values as,

$$\rho = \frac{a_o^2}{8\pi G} \left(\frac{M_{pl}}{M_D}\right)^2 = 5.369 \times 10^{-10} \frac{J}{m^3} \qquad (10)$$

This is the main formula and central result of the cosmological constant, since it allows one to make a direct comparison with observations.

As a final fun comment let us, just out of curiosity, take the formula and apply it to the entire universe. Now we note that the critical energy density of the universe equals

$$\rho_{crit} = \frac{3H_o^2 c^2}{8\pi G} = \frac{a_o^2}{8\pi G}\left(\frac{M_{pl}}{M_D}\right)^2$$

From which we obtain a relation between a_o (MONDian- determined by fits to internal properties of galaxies), H_o is the hubble constant (a measure of the present-day expansion rate of the Universe) and α_e (coupling which determines the strength of the force) as,

$$a_o = cH_o\left(\frac{M_D}{M_{pl}}\right)\sqrt{3} \qquad (11)$$

This relation holds remarkably well for the experimentally verified parameters and because it is in agreement with observations, our energy density formula (10) would be applicable to the entire universe.

The cosmological constant Λ is a dimensionful parameter with units of $(length)^{-2}$. From the point of view of classical general relativity, there is no preferred choice for what the length scale defined by Λ might be.

Particle physics, however, brings a different perspective to the question. Einstein introduced a cosmological constant into his equations for General Relativity. This term acts to counteract the gravitational pull of matter, and so it has been described as an anti-gravity effect. The cosmological constant turns out to be a measure of the energy density but no one has ever calculated the cosmological constant with confidence. Previously we showed that the cosmological constant is related to the vacuum energy density by the Friedmann relationship as,

$$\Lambda = \frac{8\pi G}{c^4} \rho$$

When we substitute for (10), we get the value of the cosmological constant as,

$$\Lambda = \frac{a_o{}^2}{c^4} \left(\frac{M_{pl}}{M_D}\right)^2 = 1.11 \times 10^{-52} m^{-2} \qquad (12)$$

Combining (11) with (12) we obtain a relation between the Hubble constant, the low acceleration and the cosmological constant as

$$a_o = \frac{c^3}{H_o} \frac{M_D}{M_{pl}} \frac{\Lambda}{\sqrt{3}}$$

In a limit when, $\frac{M_D}{M_{pl}} = \frac{H_o}{c\sqrt{\Lambda}} = 0.6956$ (This value is so close to the experimentally defined density parameter values of $\Omega = 0.6889 \pm 0.056$ Planck2018), it turns out that the best fit value for MOND is closely related to the cosmological constant, $a_o = c^2\sqrt{\frac{\Lambda}{3}}$ and this value happens at the mass scale of $M_D = 1.5164 \times 10^{-8} kg$. This has not been confirmed experimentally but it points to the theoretical observation that dark matter is a particle of a mass close Planck mass relics (or primordial black holes) which remain whenever a black hole evaporates or those which where formed in the early universe after the big bang. The parameter a_0 is the acceleration scale introduced in the phenomenological fitting formula for galaxy rotation curves. a_o is the parameter that was introduced by Milogram in MOND. It is also an explanation for the phenomenological success of Milogram's fitting formula, in particular in reproducing the flattening of rotation curves, where the asymptotic velocity of the flattened galaxy rotation curve is, $v_f^4 = a_o GM$ this is known as the baryonic Tully-Fisher relation and has been well tested by observations of a very large number of spiral galaxies.

Therefore from general formulas and assumptions given above, we have provided a precise calculation of the cosmological parameters of ρ (vacuum energy density), Λ (Cosmological constant) and H_o (Hubble constant) in agreement with the Planck (2018) observations. We have assumed only one parameter M_D which we believe will be determined by experiment this year. The results of ρ and Λ only happen when $a_o = 1.2 \times 10^{-10} m/s^2$ and, $M_D = 2.76 \times 10^{-9} kg$. Because the parameter a_0 has been determined to be a best fit for galaxy rotation curves, we therefore remain to determine by experimental means the particle with a mass M_D for galaxies and galaxy clusters. Once M_D is confirmed, the equations given will prove once and for all that the postulated Dark matter hypothesis is not responsible for what happens in galaxies and galaxy clusters.

What is real? General Relativity or Quantum Gravity

In this section we want to find out the reality behind the bending of light near the Sun's surface.

The Newtonian Approach

During Newton's time, it was believed that light was made up of particles moving at a varying speed. To prove why light bends near the Sun's surface Newton had to assume that these particles had mass. For example he considered a Sun with mass M, where a particle of light with mass m from a distant star past the Sun, had to bend near the Sun's surface due to the gravitational force of attraction acting on the particle of light. Because of this, the observer at the earth's surface never saw the actual position of the star but rather the apparent position of the star at an angle θ from its original position.

Newton assumed that, the particle of light falling freely in the gravitational field of the Sun gained kinetic energy,

$$E_k = \frac{mv^2}{2}$$

Where, v was the speed of the particle of light. The potential gravitational energy that was gained by the particle was given by,

$$V_k = \frac{GMm}{R}$$

Where R was the radius of the Sun from its centre to the point where light curved. Newton assumed that deflection angle was actually the ratio of the gravitational potential to the kinetic energy of the particle of light,

$$\theta = \frac{V_r}{E_k}$$

$$\theta = \frac{2GM}{v^2 R}$$

During Newton's time, the speed of the photon (a particle of light) was not known but, today we know this value to a much greater accuracy, thanks to Maxwell and Einstein. All the parameters, from the Sun's mass to the speed of light are known to high accuracy, therefore the Newtonian deflection angle is now known to be,

$$\theta_N = \frac{2GM}{c^2 R} = 0.875 arcsec$$

The problem with the Newtonian approach is this; we now know that photons of light are massless and move at a constant speed of light c and, the Newtonian deflection angle value is not in agreement with the observations. Therefore Newton's original calculation was flawed and required another explanation.

The Einstein Approach

In the Einstein approach it was found out that, the particles of light were called photons and that these particles where massles moving at a constant speed of light $c = 3 \times 10^8 m/s$.

Einstein's theory proposes that gravity is not an actual force, but is instead a geometric distortion of spacetime not predicted by ordinary Newtonian physics. The more mass you have to produce the gravity in a body the more distortion you get, this distortion changes the trajectories of objects moving through space, and even the paths of light rays, as they pass close-by the massive body. Even so, this effect is very feeble for an object as massive as our own sun, so it takes enormous care to even detect that it is occurring.

The Einstein deflection angle was twice the Newton's angle of deflection $\theta = 2\theta_N$, but there is no any account in literature where it shows the derivation of this deflection angle from Einstein field equations, which means that, Einstein came up with a formula similar to the Newtonian deflection angle formula given as,

$$\theta_E = \frac{4GM}{c^2 R} = 1.75 arcsec$$

Instead of the number, **2** in the Newton formula, we have a **4** and the varying speed of light in the Newton approach is replaced by a constant speed of light c.

The Einstein value was determined by observation through observing the solar eclipse. Although they say it agrees with experiment, we know that this is not true. It has long been suspected that the deflection of light in the vicinity of the sun exceeds the general relativistic predicted value of 1.75". An example of this, is the Erwin Finlay Freundlich 1929 solar eclipse expedition which produced a value of 2.24" larger than the general relativistic value. It is expected that once the reason for the deviation in the deflection angle has been found, it will disprove Einstein's imaginations for the curvature of space time.

It's almost hundred years since Sir Arthur Eddington experimentally proved Einstein's general relativity theory right. Since then, there has never been any competing theory that would prove Einstein wrong save for Loop quantum gravity and string theory. The fact that starlight is bent at the surface of the gravitating body by a deflection angle of 1.75" imposes a bound on the theoretical justification of gravity. Calculating an angle below or above 1.75" will be an upheaval in the founding blocks of physics. Erwin Finlay Freundlich was one of those people who stood out of the ordinary in 1929 when he published results with a larger angle of deflection than Eddington's. An account on Freundlich 1929 expedition has been clearly given in Robert J.Trumpler and Klaus Hentschel papers as stated below;

"Among the various expeditions sent out to observe the total solar eclipse of May 9, 1929, that of the Potsdam Observatory (Einstein Stiftung) seems to be the only one which obtained photographs suitable for determining the light deflection in the Sun's gravitational field. Two instruments were used, but so far only the results of the larger one, a 28-foot horizontal camera combined with a coelostat, have been published. The three observers, Freundlich, von Klüber, and von Brunn, claim that these observations (four plates containing from seventeen to eighteen star images each) lead to a value of 2.24" for the deflection of a light ray grazing the Sun's edge; a figure that deviates considerably from the results of the 1922 eclipse, and which is in contradiction to Einstein's generalized theory of relativity".

The irreducible anomaly in the observations of the deflection of light by the sun has been known to exist since the birth of Einstein General relativity theory. For example, in a 1959 classical review by A.A.Mikhailov, it concludes that observations yield instead of a general relativistic prediction of 1.75arcsec at the limb of the sun the simple mean value of 2.03 ± 0.10 over the GR prediction

The existence of a 2.24" deflection angle by Freundlich, Von Kluber and Von Brunn therefore implies a requirement for the modification of the general theory of relativity. Science has evolved in this simpler manner of modifications although there are some who cling to the old thoughts of "The earth is the center of the universe and Einstein is always right". I am not proving anyone wrong but I want you to believe that the general relativity theory that was put forward by Einstein is not the only 'there is' excellent description of the universe, there are other ways far better than GR as it was with the Newtonian Gravitational force replacement with a curvature of space time.

The introduction of a number **4** in Einstein deflection angle of light has no basis as to how it came along. The fact that his formula resembles the Newton formula actually shows that Einstein borrowed ideas from Newton analysis. He Einstein also failed to eliminate the mass of a photon from his equations. Even today no one knows how to deduce the deflection angle without taking into account the photon mass because we know the photon is massless.

Ladies and gentlemen, let me present to you another approach that will lead us to the Einstein deflection angle without assuming that the photon has mass or kinetic energy.

Loop Quantum Gravity Approach

Let the potential energy of the Photon according to Einstein –Planck relation be,

$$E = \frac{hc}{\lambda}$$

Where λ is the wavelength

Since ligth appears curved at a small part of the Sun's surface, then the circumference according to deBrogile is quantized in units, $C=\pi R=\lambda$ (In case light orbits the Sun, then $C=2\pi R$). Then the energy of the photon will be given by

$$E_r = \frac{2\hbar c}{R}$$

According to relativistic quantum mechanics, a photon of momentum P, has a kinetic- energy given by, where M is the Sun's mass

$$E_B = \frac{P^2}{2M}$$

According to quantum mechanics in curved space time, space is divided into small chuncks of matter (atoms of space) with a length close to the Planck length l_p, therefore the momentum of a photon passing through these atoms of space will be given by,

$$P = \frac{\hbar}{l_p}$$

This momentum is proof that the photon has no mass and what we percieve as the heaviness of the photon is actually the discrete nature of space.

Due to the discrete nature of space, there is a delay in time at which the photon will reach our telescopes from the distant star. In other words the speed of light doesn't change but there is a huge difference from the calculated time and the observed time of reach of light from the distant star. Then the energy carried by a photon through the discrete space is given as

$$E_B = \frac{\hbar c^3}{2GM}$$

This then brings us to the deflection angle which is the ratio of the photon potential energy to the kinetic energy,

$$\theta = \frac{E_r}{E_B}$$

$$\theta = \frac{4GM}{c^2 R}$$

The Extra Dimension Approach

In higher dimensions or extra dimension problems we get a different picture of what general relativity really is. We assume that light behaves differently in various dimensions and the observations of light from a distant star will vary according to the flux in the extra dimensions because it is this loss of flux to the extra dimensions which makes gravity weak yet it is strong. Therefore what determines our observations is the flux in the extra dimensions as expressed in our model below,

Let the deflection angle of light at the sun's limb be given by,

$$\theta = \frac{1}{\alpha^{n/2}}\left(\frac{R_s}{R}\right) \qquad (13)$$

Where, $R_s = \frac{2GM}{c^2}$ is the Schwarzschild radius of a gravitating body, α is the size of the extra dimension and $\alpha^{n/2}$ is the flux in the extra dimension. In what follows, we use the above equation by subsitituting in the values of $\alpha^{n/2}$ to get the values of the three deflection angles whose sample mean gives the Einstein deflection value. This analysis will help us recover new theories based on the flux in the extra dimension.

Let us start with the Newton's theory of gravitation. To recover the Newtonian deflection angle at the suns limb, we set $\alpha^{n/2} = 1$. This then gives the Newtonian value as,

$$\theta_N = \frac{R_s}{R_\odot} = 0.875 \text{arcsec}$$

The Freundlich deflection angle might have taken a different twist than with Eddington 1.75arcsec result, which we are yet to find out. Taking, $\alpha^{n/2} = 0.0233$, we deduce the deflection angle given by,

$$\theta_F = \frac{2.56 R_s}{R_\odot} = 2.24 \text{arcsec}$$

Lastly when $\alpha^{n/2} = 0.0290$ we get the following deflection angle,

$$\theta_Q = \frac{2.426 R_s}{R_\odot} = 2.12 \text{arcsec}$$

Our first result from the above calculations is that; the sample mean of the deflection angles from the three observations gives the exact deflection angle that was calculated and observed by Eddington in General relativity as,

$$\frac{\sum_{n=1}^{4} \theta_n}{3} = \frac{0.875 + 2.24 + 2.12}{3} = 1.75 arcsec$$

The fact that the mean of the three observations for the deflection of light given above reproduces the GR value of 1.75arcsec sums up what exactly general relativity really is. In simple terms GR is the sample mean of three observations taken from different location on the earth's surface where the flux in the extra dimension makes the strength of gravity slightly different in those positions where light bends.

The model given above is proof that the curvature of space assumption given by General Relatitiy was just a mathematical artifact and not a real entity. The observed deflection angles are greatly determined by the flux in the extra dimensions.

Derivation of the Energy density stored in the Electric field and Gravitational Field

We know that the electric field store energy, and that in a vacuum the energy density is given by, $\rho = \frac{\varepsilon_o}{2} E^2$ where E is the electric Field and ε_o the permittivity of free space. If our new formula for the enegy density given in the first section of this book (Eqn1) is true, it must be able to reproduce the expression for the energy density of the electric field and also solve other problems.

To derive the energy density in the electric field, we let the force on the particle say an electron with charge e due to the electric field E created by another charged electron be, F=eE. Then the energy density will be related to the electric field by,

$$\rho = \frac{e^2 E^2}{8\pi \alpha \hbar c}$$

But because the coupling constant of the electromagnetic force is the fine structure constant $\alpha = \frac{e^2}{4\pi \varepsilon_o \hbar c}$, then on substitution and cancelling like terms, we recover the energy density in the electric field as,

$$\rho = \frac{\varepsilon_0}{2} E^2$$

Similary, for the energy density in the gravitational field, let the force experienced by a particle of mass m due to the gravitational field g be F=mg. The energy density is here given by,

$$\rho = \frac{m^2 g^2}{8\pi\alpha\hbar c}$$

But because the coupling constant of the gravitational force is the fine structure constant- $\alpha = \frac{Gm^2}{\hbar c}$, then on substitution and cancelling like terms, we recover the energy density in the gravitational field as,

$$\rho = \frac{g^2}{8\pi G}$$

We have shown that, just as the electromagnetic field stores energy, the same is also true for the gravitational field.

Emergence of Gravity

Introduction

Consider two identical charged spherical bodies of mass m separated by a distance R in a vacuum under the influence of an electric field E and in presence of an external constant and uniform magnetic field B. Let one of the bodies be stationary or rigid and the other free to move. Due to the presence of an external magnetic field the charged body will experience circular motion where the surface area of the sphere created by this motion is, $A = 4\pi R^2$. After a lengthy but straightforward calculation it can be shown that the electric force Ee per unit area A on the system of bodies with charge e is given by

$$\frac{Ee}{4\pi R^2} = \frac{F_b{}^2}{4\pi \alpha \hbar c} \qquad (1)$$

Where F_b is the magnetic force experienced by a charged body, α is the fine structure constant, $\hbar$ is the reduced Planck constant and c is the constant speed of light.

The magnetic force in this case is perpendicular to the velocity v of the body as

$$F_b = Bev$$

On substitution into (1) we obtain the square of the velocity as

$$v^2 = \alpha \left(\frac{\hbar c}{R^2}\right) \frac{E}{B^2 e} \qquad (2)$$

Keeping the radius of orbit constant, it can be observed from (2) that the weaker the strength of the external magnetic field the greater the velocity and vice versa. Therefore the velocity of a body will depend on a great deal to the coupling constant or fine structure constant.

We notice that the expression $\frac{\hbar c}{R^2}$ is the Casmir force F_{cas} due to the Casmir effect. Multiplying both sides of the equation by $\frac{m}{R}$ we get an expression for the centripetal force which acts to keep the body in orbit as,

$$\frac{mv^2}{R} = \alpha F_{cas} \frac{mE}{B^2 eR}$$

According to Maxwell electromagnetic theory, the electric field is related to the magnetic field by a relation E=Bc and since $\alpha = \dfrac{e^2}{4\pi\varepsilon_0 \hbar c}$ we obtain the Casmir force under the influence of an external magnetic field between the two bodies as,

$$F = F_{cas} \dfrac{me}{4\pi R \varepsilon_0 \hbar B} \quad (3)$$

From the uncertainty principle the following relation holds,

$$R = \dfrac{\hbar}{mc}$$

Where R is the position of the mass m, which means that the circumference takes on discrete values of the wavelength as $2\pi R=\lambda$. Then on putting in (3) we obtain the force as

$$F = F_{cas} \dfrac{Mmec}{4\pi\varepsilon_0 \hbar^2 B}$$

On arranging we obtain the following expression,

$$F = \left(\frac{\hbar c}{R^2}\right)\frac{B_o}{B} \qquad (4)$$

Where B_o is the internal magnetic field between the bodies and is given as

$$B_o = \frac{m^2 ec}{4\pi\varepsilon_o \hbar^2} \qquad (5)$$

Equation 4 reduces to the gravitational force when the external magnetic field is a constant with a value given by,

$$B = \frac{ec^2}{4\pi\varepsilon_o G\hbar} = 1.8423 \times 10^{52} G \qquad (6)$$

Finally on substitution of (6) into (4) we obtain the universal law of gravitation as

$$F = \frac{Gm^2}{R^2}$$

Similarly the electrostatic force or the Coulomb force law between two electrons can be deduced in a similar manner. In this case the external magnetic field has the value of the Schwinger magnetic induction limit of

$$B = \frac{m^2 c^2}{\hbar e} = 4.3697 \times 10^{13} G \qquad (7)$$

Also the internal magnetic field (5) between two electrons is

$$B_o = \frac{m^2 ec}{4\pi \varepsilon_o \hbar^2} = 3.22 \times 10^{11} G \qquad (8)$$

Finally on substitution of (7) and (8) into (4) we get the Coulomb force law of electro statics as

$$F = \frac{e^2}{4\pi \varepsilon_o R^2}$$

While this hypothesis awaits experimental observation, it remains evident that gravitation is a Casmir effect under the influence of a constant and uniform external magnetic field of $B = 1.8423 \times 10^{52} G$.

Determining the length scale at which the force of Gravity is strong between any two electrons

Hypothesis

To begin with, let there be two identical charged particles of mass m, separated by a constant distance R in a vacuum. Let the total quantum energy stored between the particles be given by

$$W_Q = \sqrt{\alpha F_e \hbar c} \qquad (1)$$

Where, F_e is the electric force between the two particles, α is the electromagnetic coupling constant which determines the strength of the electromagnetic force, $\hbar$ is the reduced Planck constant and c is the constant speed of light.

To determine the length scale l_s at which gravity begins to act, we have to move one particle toward the other until the strength of gravity is achieved. We can do this several times until the length scale is fixed. This however becomes easier with the introduction of the magnetic field in vicinity of the particles. Assuming that one particle moves at constant speed v towards the other in the magnetic field B, then the work that will be done by one particle to move through a distance R towards the other will be given by,

$$W_B = BevR \qquad (2)$$

When this happens, then the total energy stored in the system will be in equal proportion to the work done due to the magnetic field, that is $W_Q = W_B$. Therefore equating (1) to (2) we obtain a relation between the Casimir force and velocity as

$$F_c = \frac{\hbar c}{R^2} = \frac{B^2 v^2 e}{\alpha E} \qquad (3)$$

The expression (3) implies that the Casimir force will increase as the square of the velocity with which a particle moves towards the other.

However for circular motion to take place, the centripetal force is related to the Casimir force as

$$\frac{mv^2}{R} = \alpha F_c \frac{mE}{B^2 eR}$$

In a limit when, $R = \frac{\hbar}{mc}$ (the Compton wavelength), $\alpha = \frac{e^2}{4\pi\varepsilon_0 \hbar c}$ and $E = Bc$, we determine the force of attraction between the two particles to be

$$F = F_c \frac{m^2 e c^2}{4\pi\varepsilon_0 \hbar^2 E} \qquad (4)$$

When one particle moves towards the other, we can easily determine the length at which other forces begin become strong and the electric field used to accelerate one electron towards the other in this case will follow an inverse square law in l_s^2 as

$$E = \frac{e}{4\pi\varepsilon_0 l_s^2}$$

Putting this into (4) we obtain

$$F = \left(\frac{\hbar c}{R^2}\right)\frac{l_s^2}{l_0^2} \qquad (5)$$

Where $l_o = \dfrac{\hbar}{mc}$ is taken here to be a limiting length for microscopic particles above which the descriptions of quantum mechanics cease to exist.

From equation (5) it is established that gravity will become strong when l_s is the Planck length scale given by $l_s = \sqrt{\dfrac{\hbar G}{c^3}}$, on putting this length into our formula (5) we obtain the *Newton law of gravitation* as

$$F = \frac{Gm^2}{R^2}$$

Note that, the Planck length scale is obtained by equating my electric field $E_b = \dfrac{ec^3}{4\pi\varepsilon_0 G\hbar}$ to $E = \dfrac{e}{4\pi\varepsilon_0 l_s^2}$. (see chapter6) E_b in this case is the electric field used to accelerate one electron towards the other.

Also at a length scale of $l_s = \dfrac{e}{m}\sqrt{\dfrac{\hbar}{4\pi\varepsilon_0 c^3}}$, we deduce the *Coulomb force law* of electro statistics as

$$F = \frac{e^2}{4\pi\varepsilon_0 R^2}$$

Note that, this length scale is obtained by equating the Schwinger electric field $E_s = \frac{m^2c^3}{\hbar e}$ *to* $E = \frac{e}{4\pi\varepsilon_0 l_s^2}$.

Finally, the *Casimir force* is deduced when the length scale is the Compton wavelength $l_o = \frac{\hbar}{mc}$

$$F = \frac{\hbar c}{R^2}$$

Note that, the Compton wavelength length scale is obtained by equating my electric field $E_s = \frac{m^2 ec^2}{4\pi\varepsilon_0 \hbar^2}$ *to* $E = \frac{e}{4\pi\varepsilon_0 l_s^2}$. *(see chapter6)*

While this hypothesis awaits experimental verification, it remains evident from the derivations given above that the force of gravity between two particles say electrons in an atom occurs only when the distance between them is the Planck length $l_p = 1.6144 \times 10^{-35} m$ and the magnetic field or electric field required to accelerate one particle close to the other up to the Planck length apart should be $B_b = \frac{ec^2}{4\pi\varepsilon_0 G\hbar} = 1.8423 \times 10^{52} G$ and $E_b = \frac{ec^3}{4\pi\varepsilon_0 G\hbar} = 5.5269 \times 10^{60} N/C$.

The table below shows the length at which the gravitational, electrostatic and Casimir force occur between two electrons and the strength of the Electric field required to accelerate one electron to another in order to obtain this length scale.

Force (N)	Length (m)	Electric Field (N/C)	Remarks/limit
Gravitational	1.6144×10^{-35}	5.5269×10^{60}	Quantum Gravity
Electromagnetic	3.296×10^{-14}	1.3109×10^{22}	QED limit
Casimir	3.8586×10^{-13}	9.66×10^{19}	Quantum mechanics

Revised Gravitation Theory for the Modified Newtonian Dynamics (MOND) Paradigm and Beyond

Introduction

It must be noted that the Milgrom Hypothesis that was introduced in 1982 to account for the flat rotation curves of spiral galaxies under the assumption of small acceleration fails to deduce the temperature and entropy of a black hole. In my opinion I think that all alternatives to Newton's gravity or General relativity should at least be able to explain details near or beyond black hole singularities if at all they exist. Failure for MOND to deduce the Hawking laws for black holes is one of those major indicators that it is not a genuine alternative to GR. Even if Black holes haven't been observed, I think it is in order for MOND to be consistent with the other laws of physics.

The Modified Force Law

The New force law is

$$F = \frac{me}{R}\sqrt{\frac{GM\omega}{4\pi\hbar\varepsilon_o}} \tag{1}$$

Where $\omega=2\pi f$ is the angular frequency of the graviton-photon oscillations and e is the charge on an electron. The above given law was used in chapter 10 and chapter 11 of the book "Quantum Gravity in a Nutshell1"

In a limit of $\omega = \frac{GM}{R^2}\left(\frac{4\pi\hbar\varepsilon_o}{e^2}\right) = \frac{g_N}{c\alpha_e}$, where g_N is the usual Newtonian acceleration due to gravity and α_e is the fine structure constant, the above new force law reduces to the Newtonian law of universal gravitation.

The Tully-Fisher Relation

One of the best fit predictions of MOND is a single universal Tully-Fisher relation.

" The relation between asymptotic velocity and the mass of the galaxy is an absolute one" (Milgrom 1983). This is given by, $V^4 = a_o GM$, where $a_o = 1.2 \times 10^{-10} ms^{-2}$. In this paper an equation similar to the Tully-Fisher relation is deduced from (1) as given below,

For circular orbits about a point mass, M

$$\frac{V^2}{R} = \frac{e}{R}\sqrt{\frac{GM\omega}{4\pi\hbar\varepsilon_o}}$$

This gives an asymptotically rotation velocity independent of R:

$$V^4 = \left(\frac{e^2\omega}{4\pi\hbar\varepsilon_o}\right) GM = \omega c\alpha_e GM \qquad (2)$$

It is this behavior that gives rise to asymptotically flat rotation curves and the Tully-Fisher relation (Tully & Fisher 1977)

Comparing (2) to the Tully-Fisher relation, we determine the angular frequency of the graviton or photon oscillation as,

$$\omega = \frac{a_o}{c\alpha_e} = 5.48 \times 10^{-17} rad/s$$

Graviton-Photon energy

It has been known that the energy of a photon is related to the angular frequency by, $E = \hbar\omega$. This equation worked well at the atomic scale but it could not explain the spectrum of the galaxy clusters. Here we give a formula that works at the galactic scale as given below,

From (2) $\omega = \dfrac{V^4}{c\alpha_e GM}$, then

$$E = \hbar\omega = \dfrac{\hbar V^4}{c\alpha_e GM} \qquad (3)$$

Comparing this to the total energy of the gravitating body from Einstein Energy relation, we have

$$Mc^2 = \dfrac{\hbar V^4}{c\alpha_e GM}$$

This then gives,

$$\dfrac{V^4}{c^4} = \alpha_e \alpha_g$$

Where, $\alpha_g = \dfrac{GM^2}{\hbar c}$ is the gravitational coupling constant

This shows that the velocity of the stars in circular orbit about a center of mass M will never exceed that of light by, $V = c(\alpha_e \alpha_g)^{1/4}$.

Derivation of the Temperature of a Black Hole from the new force law

A Black hole evaporates this way; the gravitational force of the entire mass of a black hole balances the electromagnetic forces between individual electrons inside the Black hole. When this happens, electrons inside a black hole acquire the same charge sign as with those at the surface and therefore a particle is emitted from the surfaces of a black hole

$$\frac{e^2}{4\pi\hbar\varepsilon_o R^2} = \frac{me}{R}\sqrt{\frac{GM\omega}{4\pi\hbar\varepsilon_o}}$$

Squaring both sides of the above equation and arranging, in a limit v=c, we have the kinetic energy of the emitted particle as,

$$KE = mc^2 = \frac{4e^2 \lambda \mu_o \hbar c^3}{8\pi GmMA}$$

Where λ is the wavelength of a particle emitted from a black hole, μ_o is the permeability of free space and

$A = 4\pi R^2$ is the surface area of the event Horizon of a Black hole.

By equivalence, $kT = \dfrac{4e^2 \lambda \square_o \hbar c^3}{8\pi GmMA}$

Where kT is the thermal kinetic energy of the emitted particle, then the temperature of the emitted particle is,

$$T = \dfrac{4e^2 \lambda \square_o}{mA}\left(\dfrac{\hbar c^3}{8\pi GMk}\right)$$

In a limit, $= \dfrac{mA}{4e^2 \square_o}$, the above equation gives the temperature of a black hole as,

$$T = \left(\dfrac{\hbar c^3}{8\pi GMk}\right)$$

In conclusion, the Milogram Hypothesis that works well on a classical level fails to work on a quatum level. In this kind of situation a theory as given above is required to explain details where the quantum gravitation effects become important. This still points to the requirement of a quantum theory of gravity. In this paper, we have deduced a near approximation to the quantum theory of gravity which works well in explaining the spiral galaxy rotation curves and the energy associated with the quanta emitted in the process.

Epilogue

I hope that this book has succeeded in describing to you, the reader, how difficult it is to try to wrest precious, fundamental secrets from nature. That quest can be compared to climbing a mountain, and when reaching the peak, seeing another higher mountain that tempts us to ascend to even greater heights. And when we do reach the higher peak, we discover as we look across the valley yet another peak that calls. In the end, it is the wonderful experience of scaling the mountain—of attempting to understand the secrets of nature—that motivates us as scientists. There is of course the additional thrill, upon reaching the top of a mountain, to ram in the flagpole announcing one's victory. But that is only a momentary emotion soon superseded by the new challenges presented by the higher peak on the horizon.

For readers who have a background in physics, let us think about Einstein's gravitational theory in an abstract, whole sense. Then let us think about QG. Do we get the same aesthetic pleasure from considering QG as we do from Einstein's gravity theory? To reach a true appreciation of the elegance of QG as a gravity theory, it is necessary to explore all the technical details of QG and see how it works as a whole theoretical framework. One has to experience its successes in explaining data and naturally allowing for a cosmology with no singularity at the beginning of the universe, and no dark matter, and a unified description of the accelerating universe. Only after the laborious work of achieving a technical understanding of QG's theoretical structure can one truly appreciate QG's elegance. I hope that future generations of physicists will be motivated to study the theory in the same depth as Carlo Rovelli, Joel Brownstein, and I have done, and appreciate its intrinsic beauty.

There is still important research to perform before we have a complete picture of where we stand with quantum gravity. Perhaps with more attention being paid by other physicists who can investigate QG and apply it to other observational data, we will arrive at a more convincing state-of-the-art of gravity. The ultimate tests for QG, or any alternative gravity theory, can be stated simply: With a minimum number of assumptions that are physically consistent, how much observational data can be explained? More important, can the theory make testable predictions that cannot be accounted for by competing theories? In the latter part of this book, I have suggested several ways that future observations and experiments can verify or falsify QG.

In probing the mysteries of nature, physicists need to have faith that we can through mathematical equations reach a true understanding of nature such that the predictions of our equations can be verified by experiment or observation. We need to continually aspire to that goal despite the modern trend in theoretical physics of indulging in speculations that can never be proved or falsified by reality.

Glossary

Absolute space and time—the Newtonian concepts of space and time, in which space is independent of the material bodies within it, and time flows at the same rate throughout the universe without regard to the locations of different observers and their experience of "now."

Acceleration—the rate at which the speed or velocity of a body changes.

Accelerating universe—the discovery in 1998, through data from very distant supernovae, that the expansion of the universe in the wake of the big bang is not slowing down, but is actually speeding up at this point in its history; groups of astronomers in California and Australia independently discovered that the light from the supernovae appears dimmer than would be expected if the universe were slowing down.

Action—the mathematical expression used to describe a physical system by requiring only the knowledge of the initial and final states of the system; the values of the physical variables at all intermediate states are determined by minimizing the action.

Anthropic principle—the idea that our existence in the universe imposes constraints on its properties; an extreme version claims that we owe our existence to this principle.

Asymptotic freedom (or safety)—a property of quantum field theory in which the strength of the coupling between elementary particles vanishes with increasing energy and/or decreasing distance, such that the elementary particles approach free particles with no external forces acting on them; moreover for decreasing energy and/or increasing distance between the particles, the strength of the particle force increases indefinitely.

Baryon—a subatomic particle composed of three quarks, such as the proton and neutron.

Big bang theory—the theory that the universe began with a violent explosion of spacetime, and that matter and energy originated from an infinitely small and dense point.

Big crunch—similar to the big bang, this idea postulates an end to the universe in a singularity.

Binary stars—a common astrophysical system in which two stars rotate around each other; also called a "double star."

Blackbody—a physical system that absorbs all radiation that hits it, and emits characteristic radiation energy depending upon temperature; the concept of blackbodies is useful, among other things, in learning the temperature of stars.

Black hole—created when a dying star collapses to a singular point, concealed by an "event horizon;" the black hole is so dense and has such strong gravity that nothing, including light, can escape it; black holes are predicted by general relativity, and though they cannot be "seen," several have been inferred from astronomical observations of binary stars and massive collapsed stars at the centers of galaxies.

Boson—a particle with integer spin, such as photons, mesons, and gravitons, which carries the forces between fermions.

Brane—shortened from "membrane," a higher-dimensional extension of a onedimensional string.

Cassini spacecraft—NASA mission to Saturn, launched in 1997, that in addition to making detailed studies of Saturn and its moons, determined a bound on the variations of Newton's gravitational constant with time.

Causality—the concept that every event has in its past events that caused it, but no event can play a role in causing events in its past.

Classical theory—a physical theory, such as Newton's gravity theory or Einstein's general relativity, that is concerned with the macroscopic universe, as opposed to theories concerning events at the submicroscopic level such as quantum mechanics and the standard model of particle physics.

Copernican revolution—the paradigm shift begun by Nicolaus Copernicus in the early sixteenth century, when he identified the sun, rather than the Earth, as the center of the known universe.

Cosmic microwave background (CMB)—the first significant evidence for the big bang theory; initially found in 1964 and studied further by NASA teams in 1989 and the early 2000s, the CMB is a smooth signature of microwaves everywhere in the sky, representing the "afterglow" of the big bang: Infrared light produced about 400,000 years after the big bang had redshifted through the stretching of spacetime during fourteen billion years of expansion to the microwave part of the electromagnetic spectrum, revealing a great deal of information about the early universe.

Cosmological constant—a mathematical term that Einstein inserted into his gravity field equations in 1917 to keep the universe static and eternal; although he later regretted this and called it his "biggest blunder," cosmologists today still use the

cosmological constant, and some equate it with the mysterious dark energy.

Coupling constant—a term that indicates the strength of an interaction between particles or fields; electric charge and Newton's gravitational constant are coupling constants.

Crystalline spheres—concentric transparent spheres in ancient Greek cosmology that held the moon, sun, planets, and stars in place and made them revolve around the Earth; they were part of the western conception of the universe until the Renaissance.

Curvature—the deviation from a Euclidean spacetime due to the warping of the geometry by massive bodies.

Dark energy—a mysterious form of energy that has been associated with negative pressure vacuum energy and Einstein's cosmological constant; it is hypothesized to explain the data on the accelerating expansion of the universe; according to the standard model, the dark energy, which is spread uniformly

throughout the universe, makes up about 70 percent of the total mass and energy content of the universe.

Dark matter—invisible, not-yet-detected, unknown particles of matter, representing about 30 percent of the total mass of matter according to the standard model; its presence is necessary if Newton's and Einstein's gravity theories are to fit data from galaxies, clusters of galaxies, and cosmology; together, dark

matter and dark energy mean that 96 percent of the matter and energy in the universe is invisible.

Deferent—in the ancient Ptolemaic concept of the universe, a large circle representing the orbit of a planet around the Earth.

Doppler principle or **Doppler effect**—the discovery by the nineteenth-century Austrian scientist Christian Doppler that when sound or light waves are moving toward an observer, the apparent frequency of the waves will be shortened, while if they are moving away from an observer, they will be lengthened; in

astronomy this means that the light emitted by galaxies moving away from us is redshifted, and that from nearby galaxies moving toward us is blueshifted.

Dwarf galaxy—a small galaxy (containing several billion stars) orbiting a larger galaxy; the Milky Way has over a dozen dwarf galaxies as companions, including the Large Magellanic Cloud and Small Magellanic Cloud.

Dynamics—the physics of matter in motion.

Electromagnetism—the unified force of electricity and magnetism, discovered to be the same phenomenon by Michael Faraday and James Clerk Maxwell in the nineteenth century.

Electromagnetic radiation—a term for wave motion of electromagnetic fields which propagate with the speed of light—300,000 kilometers per second—and differ only in wavelength; this includes visible light, ultraviolet light, infrared radiation,

X-rays, gamma rays, and radio waves.

Electron—an elementary particle carrying negative charge that orbits the nucleus of an atom.

Eötvös experiments—torsion balance experiments performed by Hungarian Count Roland von Eötvös in the late nineteenth and early twentieth centuries that showed that inertial and gravitational mass were the same to one part in 1011; this was a more accurate determination of the equivalence principle than results achieved by Isaac Newton and, later, Friedrich Wilhelm Bessel.

Epicycle—in the Ptolemaic universe, a pattern of small circles traced out by a planet at the edge of its "deferent" as it orbited the Earth; this was how the Greeks accounted for the apparent retrograde motions of the planets.

Equivalence principle—the phenomenon first noted by Galileo that bodies falling in a gravitational field fall at the same rate, independent of their weight and composition; Einstein extended the principle to show that gravitation is identical (equivalent) to acceleration.

Escape velocity—the speed at which a body must travel in order to escape a strong gravitational field; rockets fired into orbits around the Earth have calculated escape velocities, as do galaxies at the periphery of galaxy clusters.

Ether (or aether)—a substance whose origins were in the Greek concept of "quintessence," the ether was the medium through which energy and matter moved, something more than a vacuum and less than air; in the late nineteenth century the Michelson-Morley experiment disproved the existence of the ether.

Euclidean geometry—plane geometry developed by the third-century bc Greek mathematician Euclid; in this geometry, parallel lines never meet.

Fermion—a particle with half-integer spin, like protons and electrons, that make up matter.

Field—a physical term describing the forces between massive bodies in gravity and electric charges in electromagnetism; Michael Faraday discovered the concept of field when studying magnetic conductors.

Field equations—differential equations describing the physical properties of interacting massive particles in gravity and electric charges in electromagnetism; Maxwell's equations for electromagnetism and Einstein's equations of gravity are prominent examples in physics.

Fifth force or **"skew" force**—a new force in MOG that has the effect of modifying gravity over limited length scales; it is carried by a particle with mass called the phion.

Fine-tuning—the unnatural cancellation of two or more large numbers involving an absurd number of decimal places, when one is attempting to explain a physical phenomenon; this signals that a true understanding of the physical phenomenon has not been achieved.

Fixed stars—an ancient Greek concept in which all the stars were static in the sky, and moved around the Earth on a distant crystalline sphere.

Frame of reference—the three spatial coordinates and one time coordinate that an observer uses to denote the position of a particle in space and time.

Galaxy—organized group of hundreds of billions of stars, such as our Milky Way.

Galaxy cluster—many galaxies held together by mutual gravity but not in as organized a fashion as stars within a single galaxy.

Galaxy rotation curve—a plot of the Doppler shift data recording the observed velocities of stars in galaxies; those stars at the periphery of giant spiral galaxies are observed to be going faster than they "should be" according to Newton's and Einstein's gravity theories.

General relativity—Einstein's revolutionary gravity theory, created in 1916 from a mathematical generalization of his theory of special relativity; it changed our concept of gravity from Newton's universal force to the warping of the geometry of spacetime in the presence of matter and energy.

Geodesic—the shortest path between two neighboring points, which is a straight line in Euclidian geometry, and a unique curved path in four-dimensional spacetime.

Globular cluster—a relatively small, dense system of up to millions of stars occurring commonly in galaxies.

Gravitational lensing—the bending of light by the curvature of spacetime; galaxies and clusters of galaxies act as lenses, distorting the images of distant bright galaxies or quasars as the light passes through or near them.

Gravitational mass—the active mass of a body that produces a gravitational force on other bodies.

Gravitational waves—ripples in the curvature of spacetime predicted by general relativity; although any accelerating body can produce gravitational radiation or waves, those that could be detected by experiments would be caused by cataclysmic cosmic events.

Graviton—the hypothetical smallest packet of gravitational energy, comparable to the photon for electromagnetic energy; the graviton has not yet been seen experimentally.

Group (in mathematics)—in abstract algebra, a set that obeys a binary operation that satisfies certain axioms; for example, the property of addition of integers makes a group; the branch of mathematics that studies groups is called group theory.

Hadron—a generic word for fermion particles that undergo strong nuclear interactions.

Hamiltonian—an alternative way of deriving the differential equations of motion for a physical system using the calculus of variations; Hamilton's principle is also called the "principle of stationary action" and was originally formulated by Sir William Rowan Hamilton for classical mechanics; the principle applies to classical fields such as the gravitational and electromagnetic fields, and has had important applications in quantum mechanics and quantum field theory.

Homogeneous—in cosmology, when the universe appears the same to all observers, no matter where they are in the universe.

Inertia—the tendency of a body to remain in uniform motion once it is moving, and to stay at rest if it is at rest; Galileo discovered the law of inertia in the early seventeenth century.

Inertial mass—the mass of a body that resists an external force; since Newton, it has been known experimentally that inertial and gravitational mass are equal; Einstein used this equivalence of inertial and gravitational mass to postulate his equivalence principle, which was a cornerstone of his gravity theory.

Inflation theory—a theory proposed by Alan Guth and others to resolve the flatness, horizon, and homogeneity problems in the standard big bang model; the very early universe is pictured as expanding exponentially fast in a fraction of a second.

Interferometry—the use of two or more telescopes, which in combination create a receiver in effect as large as the distance between them; radio astronomy in particular makes use of interferometry.

Inverse square law—discovered by Newton, based on earlier work by Kepler, this law states that the force of gravity between two massive bodies or point particles decreases as the inverse square of the distance between them.

Isotropic—in cosmology, when the universe looks the same to an observer, no matter in which direction she looks.

Kelvin temperature scale—designed by Lord Kelvin (William Thomson) in the mid-1800s to measure very cold temperatures, its starting point is absolute zero, the coldest possible temperature in the universe, corresponding to −273.15 degrees Celsius; water's freezing point is 273.15K (0°C), while its boiling point is 373.15K (100°C).

Lagrange points—discovered by the Italian-French mathematician Joseph-Louis Lagrange, these five special points are in the vicinity of two orbiting masses where a third, smaller mass can orbit at a fixed distance from the larger masses; at the Lagrange points, the gravitational pull of the two large masses precisely equals the centripetal force required to keep the third body, such as a satellite, in a bound orbit; three of the Lagrange points are unstable, two are stable.

Lagrangian—named after Joseph-Louis Lagrange, and denoted by L, this mathematical expression summarizes the dynamical properties of a physical system; it is defined in classical mechanics as the kinetic energy T minus the potential energy V; the equations of motion of a system of particles may be derived from the Euler-

Lagrange equations, a family of partial differential equations.

Light cone—a mathematical means of expressing past, present, and future space and time in terms of spacetime geometry; in four-dimensional Minkowski spacetime, the light rays emanating from or arriving at an event separate spacetime into a past cone and a future cone which meet at a point corresponding

to the event.

Lorentz transformations—

mathematical transformations from one inertial frame of reference to another such that the laws of physics remain the same; named after Hendrik Lorentz, who developed them in 1904, these transformations form the basic mathematical equations underlying special relativity.

Mercury anomaly—a phenomenon in which the perihelion of Mercury's orbit advances more rapidly than predicted by Newton's equations of gravity; when Einstein showed that his gravity theory predicted the anomalous precession, it was the first empirical evidence that general relativity might be correct.

Meson—a short-lived boson composed of a quark and an antiquark, believed to bind protons and neutrons together in the atomic nucleus.

Metric tensor—mathematical symmetric tensor coefficients that determine the infinitesimal distance between two points in spacetime; in effect the metric tensor distinguishes between Euclidean and non-Euclidean geometry.

Michelson-Morley experiment—1887 experiment by Albert Michelson and Edward Morley that proved that the ether did not exist; beams of light traveling in the same direction, and in the perpendicular direction, as the supposed ether showed no difference in speed or arrival time at their destination.

Milky Way—the spiral galaxy that contains our solar system.

Minkowski spacetime—the geometrically flat spacetime, with no gravitational effects, first described by the Swiss mathematician Hermann Minkowski; it became the setting of Einstein's theory of gravity.

MOG—my relativistic modified theory of gravitation, which generalizes Einstein's general relativity; MOG stands for "Modified Gravity."

MOND—a modification of Newtonian gravity published by Mordehai Milgrom in 1983; this is a nonrelativistic phenomenological model used to describe rotational velocity curves of galaxies; MOND stands for "Modified Newtonian

Dynamics."

Neutrino—an elementary particle with zero electric charge; very difficult to detect, it is created in radioactive decays and is able to pass through matter almost undisturbed; it is considered to have a tiny mass that has not yet been accurately measured.

Neutron—an elementary and electrically neutral particle found in the atomic nucleus, and having about the same mass as the proton.

Nuclear force—another name for the strong force that binds protons and neutrons together in the atomic nucleus.

Nucleon—a generic name for a proton or neutron within the atomic nucleus.

Neutron star—the collapsed core of a star that remains after a supernova explosion; it is extremely dense, relatively small, and composed of neutrons.

Newton's gravitational constant—the constant of proportionality, G, which occurs in the Newtonian law of gravitation, and says that the attractive force between

two bodies is proportional to the product of their masses and inversely proportional to the square of the distance between them; its numerical value is: $G = 6.67428 \pm 0.00067 \times 10^{-11} \, m^3 \, kg^{-1} \, s^{-2}$.

Nonsymmetric field theory (Einstein)—a mathematical description of the geometry of spacetime based on a metric tensor that has both a symmetric part and an antisymmetric part; Einstein used this geometry to formulate a unified field

theory of gravitation and electromagnetism.

Nonsymmetric Gravitation Theory (NGT)—my generalization of Einstein's purely gravitation theory (general relativity) that introduces the antisymmetric field as an extra component of the gravitational field; mathematically speaking, the nonsymmetric field structure is described by a non-Riemannian geometry.

Parallax—the apparent movement of a nearer object relative to a distant background when one views the object from two different positions; used with triangulation for measuring distances in astronomy.

Paradigm shift—a revolutionary change in belief, popularized by the philosopher Thomas Kuhn, in which the majority of scientists in a given field discard a traditional theory of nature in favor of a new one that passes the tests of experiment and observation; Darwin's theory of natural selection, Newton's gravity theory, and Einstein's general relativity all represented paradigm shifts.

Parsec—a unit of astronomical length equal to 3.262 light years.

Particle-wave duality—the fact that light in all parts of the electromagnetic spectrum (including radio waves, X-rays, etc., as well as visible light) sometimes acts like waves and sometimes acts like particles or photons; gravitation may be similar, manifesting as waves in spacetime or graviton particles.

Perihelion—the position in a planet's elliptical orbit when it is closest to the sun.

Perihelion advance—the movement, or changes, in the position of a planet's perihelion in successive revolutions of its orbit over time; the most dramatic perihelion advance is Mercury's, whose orbit traces a rosette pattern.

Perturbation theory—a mathematical method for finding an approximate solution to an equation that cannot be solved exactly, by expanding the solution in a series in which each successive term is smaller than the preceding one.

Phion—name given to the massive vector field in MOG; it is represented both by a boson particle, which carries the fifth force, and a field.

Photoelectric effect—the ejection of electrons from a metal by X-rays, which proved the existence of photons; Einstein's explanation of this effect in 1905 won him the Nobel Prize in 1921; separate experiments proving and demonstrating

the existence of photons were performed in 1922 by Thomas Millikan and Arthur Compton, who received the Nobel Prize for this work in 1923 and 1927, respectively.

Photon—the quantum particle that carries the energy of electromagnetic waves; the spin of the photon is 1 times Planck's constant h.

Pioneer 10 and 11 spacecraft—launched by NASA in the early 1970s to explore the outer solar system, these spacecraft show a small, anomalous acceleration as they leave the inner solar system.

Planck's constant (h)—a fundamental constant that plays a crucial role in quantum mechanics, determining the size of quantum packages of energy such as the photon; it is named after Max Planck, a founder of quantum mechanics

Principle of general covariance—Einstein's principle that the laws of physics remain the same whatever the frame of reference an observer uses to measure physical quantities.

Principle of least action—more accurately the principle of *stationary* action, this variational principle, when applied to a mechanical system or a field theory, can be used to derive the equations of motion of the system; the credit for discovering the principle is given to Pierre-Louis Moreau Maupertius but it may have been discovered independently by Leonhard Euler or Gottfried Leibniz.

Proton—an elementary particle that carries positive electrical charge and is the nucleus of a hydrogen atom.

Ptolemaic model of the universe—the predominant theory of the universe until the Renaissance, in which the Earth was the heavy center of the universe and all other heavenly bodies, including the moon, sun, planets, and stars, orbited around it; named for Claudius Ptolemy.

Quantize—to apply the principles of quantum mechanics to the behavior of matter and energy (such as the electromagnetic or gravitational field energy); breaking down a field into its smallest units or packets of energy.

Quantum field theory—the modern relativistic version of quantum mechanics used to describe the physics of elementary particles; it can also be used in nonrelativistic fieldlike systems in condensed matter physics.

Quantum gravity—attempts to unify gravity with quantum mechanics.

Quantum mechanics—the theory of the interaction between quanta (radiation) and matter; the effects of quantum mechanics become observable at the submicroscopic distance scales of atomic and particle physics, but macroscopic quantum effects can be seen in the phenomenon of quantum entanglement.

Quantum spin—the intrinsic quantum angular momentum of an elementary particle; this is in contrast to the classical orbital angular momentum of a body rotating about a point in space.

Quark—the fundamental constituent of all particles that interact through the strong nuclear force; quarks are fractionally charged, and come in several varieties; because they are confined within particles such as protons and neutrons, they cannot be detected as free particles.

Quasars—"quasi-stellar objects," the farthest distant objects that can be detected with radio and optical telescopes; they are exceedingly bright, and are believed to be young, newly forming galaxies; it was the discovery of quasars in 1960 that quashed the steady-state theory of the universe in favor of the big bang.

Quintessence—a fifth element in the ancient Greek worldview, along with earth, water, fire and air, whose purpose was to move the crystalline spheres that supported the heavenly bodies orbiting the Earth; eventually this concept became known as the "ether," which provided the *something* that bodies needed to be in contact with in order to move; although Einstein's special theory of relativity dispensed with the ether, recent explanations of the acceleration of the universe call the varying negative pressure vacuum energy "quintessence."

Redshift—a useful phenomenon based on the Doppler principle that can indicate whether and how fast bodies in the universe are receding from an observer's position on Earth; as galaxies move rapidly away from us, the frequency of the wavelength of their light is shifted toward the red end of the electromagnetic spectrum; the amount of this shifting indicates the distance of the galaxy.

Riemann curvature tensor—a mathematical term that specifies the curvature of four-dimensional spacetime.

Riemannian geometry—a non-Euclidean geometry developed in the mid-nineteenth century by the German mathematician George Bernhard Riemann that describes curved surfaces on which parallel lines *can* converge, diverge, and even intersect, unlike Euclidean geometry; Einstein made Riemannian geometry the mathematical formalism of general relativity.

Satellite galaxy—a galaxy that orbits a host galaxy or even a cluster of galaxies.

Scalar field—a physical term that associates a value without direction to every point in space, such as temperature, density, and pressure; this is in contrast to a vector field, which has a direction in space; in Newtonian physics or in electrostatics, the potential energy is a scalar field and its gradient is the vector force field; in quantum field theory, a scalar field describes a boson particle with spin zero.

Scale invariance—distribution of objects or patterns such that the same shapes and distributions remain if one increases or decreases the size of the length scales or space in which the objects are observed; a common example of scale invariance

is fractal patterns.

Schwarzschild solution—an exact spherically symmetric static solution of Einstein's field equations in general relativity, worked out by the astronomer Karl Schwarzschild in 1916, which predicted the existence of black holes.

Self-gravitating system—a group of objects or astrophysical bodies held together by mutual gravitation, such as a cluster of galaxies; this is in contrast to a "bound system" like our solar system, in which bodies are mainly attracted to and revolve around a central mass.

Singularity—a place where the solutions of differential equations break down; a spacetime singularity is a position in space where quantities used to determine the gravitational field become infinite; such quantities include the curvature of spacetime and the density of matter.

Spacetime—in relativity theory, a combination of the three dimensions of space with time into a four-dimensional geometry; first introduced into relativity by Hermann Minkowski in 1908.

Special theory of relativity—Einstein's initial theory of relativity, published in 1905, in which he explored the "special" case of transforming the laws of physics from one uniformly moving frame of reference to another; the equations of special relativity revealed that the speed of light is a constant, that objects appear contracted in the direction of motion when moving at close to the speed of light, and that $E = mc^2$, or energy is equal to mass times the speed of light squared.

Spin—see quantum spin.

String theory—a theory based on the idea that the smallest units of matter are not point particles but vibrating strings; a popular research pursuit in physics for two decades, string theory has some attractive mathematical features, but has yet to make a testable prediction.

Strong force—see nuclear force.

Supernova—spectacular, brilliant death of a star by explosion and the release of heavy elements into space; supernovae type 1a are assumed to have the same intrinsic brightness and are therefore used as standard candles in estimating cosmic distances.

Supersymmetry—a theory developed in the 1970s which, proponents claim, describes the most fundamental spacetime symmetry of particle physics: For every boson particle there is a supersymmetric fermion partner, and for every fermion there exists a supersymmetric boson partner; to date, no supersymmetric particle partner has been detected.

Tully-Fisher law—a relation stating that the asymptotically flat rotational velocity of a star in a galaxy, raised to the fourth power, is proportional to the mass or luminosity of the galaxy.

Unified theory (or unified field theory)—a theory that unites the forces of nature; in Einstein's day those forces consisted of electromagnetism and gravity; today the weak and strong nuclear forces must also be taken into account, and perhaps someday MOG's fifth force or skew force will be included; no one has yet discovered a successful unified theory.

Vacuum—in quantum mechanics, the lowest energy state, which corresponds to the vacuum state of particle physics; the vacuum in modern quantum field theory is the state of perfect balance of creation and annihilation of particles and antiparticles.

Variable Speed of Light (VSL) cosmology—an alternative to inflation theory, in which the speed of light was much faster at the beginning of the universe than it is today; like inflation, this theory solves the horizon and flatness problems in the very early universe in the standard big bang model.

Vector field—a physical value that assigns a field with the position and direction of a vector in space; it describes the force field of gravity or the electric and magnetic force fields in James Clerk Maxwell's field equations.

Virial theorem—a means of estimating the average speed of galaxies within galaxy clusters from their estimated average kinetic and potential energies.

Vulcan—a hypothetical planet predicted by the nineteenth-century astronomer Urbain Jean Joseph Le Verrier to be the closest orbiting planet to the sun; the presence of Vulcan would explain the anomalous precession of the perihelion of Mercury's orbit; Einstein later explained the anomalous precession in general relativity by gravity alone.

Weak force—one of the four fundamental forces of nature, associated with radioactivity such as beta decay in subatomic physics; it is much weaker than the strong nuclear force but still much stronger than gravity.

X-ray clusters—galaxy clusters with large amounts of extremely hot gas within them that emit X-rays; in such clusters, this hot gas represents at least twice the mass of the luminous stars.

Bibliography

Balungi Francis, (2010) "A hypothetical investigation into the realm of the microscopic and macroscopic universes beyond the standard model" general physics arXiv:1002.2287v1 [physics.gen-ph]

Hawking, Stephen (1975). "Particle Creation by Black Holes". Commun. Math. Phys. 43 (3): 199–220. Bibcode:1975CMaPh..43..199H.

Hawking, S. W. (1974). "Black hole explosions?". Nature.248(5443):30–31.

Bibcode:1974Natur.248...30H.doi:10.1038/248030a0.

Carlo Rovelli (2003) "Quantum Gravity" Draft of the Book Pdf

Some few texts used are from Wikipedia https://creativecommons.org/licenses/by-sa/3.0/

D. N. Page, Phys. Rev. D 13, 198 (1976).

C. Gao and Y.Lu, Pulsations of a black hole in LQG (2012) arXiv:1706.08009v3

A.H. Chamseddine and V.Mukhanov, Non singular black hole (2016) arXiv 1612.05861v1

M.Bojowald and G.M.Paily, A no-singularity scenario in LQG (2012) arXiv: 1206.5765v1

P.Singh, class.Quant.Grav,26,125005(2009), arXiv:0901.2750

P.Singh and F.Vidotto, Phys.Rev, D83,064027(2011) arXiv:1012.1307

C.Rovelli and F.Vidotto, Phy. Rev,111(9) 091303(2013) arXiv:1307.3228v2

M.Bojowald, Initial conditions for a universe, Gravity Research Foundation (2003)

A.Ashtekar, Singularity Resolution in Loop Quantum Cosmology (2008) arXiv:0812.4703v1

J.Brunneumann and T.Thiemann, On singularity avoidance in Loop Quantum Gravity (2005) arXiv:0505032v1

L.Modesto, Disappearence of the Black hole singularity in Quantum gravity (2004) arXiv:0407097v2

Mikhailov, A.A. (1959).Mon. Not. Roy. Astron. Soc.,119, 593.

P. Merat etal.(1974). Astron & Astrophys 32, 471-475

Trempler, R.J. (1956).Helv. Phys. Acta, Suppl.,IV, 106.

Trempler, R.J. (1932). " The deflection of light in the sun's gravitational field "Astronomical society of the pacific 167

Einstein, A. (1916).Ann. d. Phys.,49, 769; (1923).The Principle of Relativity, (translators Perret, W. and Jeffery, G.B.), (Dover Publications, Inc., New York), pp. 109–164.

Von Klüber, H. (1960). InVistas in Astronomy, Vol. 3, pp. 47–77.

K. Hentschel (1992). Erwin Finlay Freundlich and testing Einstein theory of relativity, Communicated by J.D. North

Muhleman, D.O., Ekers, R.D. and Fomalont, E.B. (1970).Phys. Rev. Lett.,24, 1377

Mikhailov, A.A. (1956).Astron. Zh.,33, 912.

Dyson, F.W., Eddington, A.S. and Davidson, C. (1920).Phil. Trans. Roy. sog., A220, 291

Chant, C.A. and Young, R.K. (1924).Publ. Dom. Astron. Obs.,2, 275.

Campbell, W.W. and Trumbler, R.J. (1923).Lick Obs. Bull.,11, 41.

Freundlich, E.F., von Klüber, H. and von Brunn, A. (1931).Abhandl. Preuss. Akad. Wiss. Berlin, Phys. Math. Kl., No.l;Z. Astrophys.,3, 171

Mikhailov, A.A. (1949).Expeditions to Observe the Total Solar Eclipse of 21 September, 1941, (report), (ed. Fesenkov, V.G.), (Publications of the Academy of Sciences, U.S.S.R.), pp. 337–351.

S.P. Martin, in Perspectives on Supersymmetry, edited by G.L. Kane (World Scientific, Singapore, 1998) pp. 1–98; and a longer archive version in hep-ph/9709356; I.J.R. Aitchison, hep-ph/0505105.

Mara Beller, Quantum Dialogue: The Making of a Revolution. University of Chicago Press, Chicago, 2001.

Morrison, Philp: "The Neutrino, scientific American, Vol 194,no.1 (1956),pp.58-68.

R. Haag, J. T. Lopuszanski and M. Sohnius, Nucl. Phys. B88, 257 (1975) S.R. Coleman and J. Mandula, Phys.Rev. 159 (1967) 1251.

H.P. Nilles, Phys. Reports 110, 1 (1984).

P. Nath, R. Arnowitt, and A.H. Chamseddine, Applied $N = 1$ Supergravity (World Scientific, Singapore, 1984).

S.P. Martin, in Perspectives on Supersymmetry , edited by G.L. Kane (World Scientific, Singapore, 1998) pp. 1–98; and a longer archive version in hep-ph/9709356; I.J.R. Aitchison, hep-ph/0505105.

S. Weinberg, The Quantum Theory of Fields, VolumeIII: Supersymmetry (Cambridge University Press, Cambridge,UK, 2000).

E. Witten, Nucl. Phys. B188, 513 (1981).

S. Dimopoulos and H. Georgi, Nucl. Phys. B193, 150(1981).

N. Sakai, Z. Phys. C11, 153 (1981);R.K. Kaul, Phys. Lett. 109B, 19 (1982).

L. Susskind, Phys. Reports 104, 181 (1984).

L. Girardello and M. Grisaru, Nucl. Phys. B194, 65(1982); L.J. Hall and L. Randall,

Phys. Rev. Lett. 65, 2939(1990);I. Jack and D.R.T. Jones, Phys. Lett. B457, 101 (1999).

For a review, see N. Polonsky, Supersymmetry: Structureand phenomena. Extensions of the standard model, Lect.Notes Phys. M68, 1 (2001).

G. Bertone, D. Hooper and J. Silk, Phys. Reports 405, 279 (2005).

G. Jungman, M. Kamionkowski, and K. Griest, Phys. Reports 267, 195 (1996).

V. Agrawal, S.M. Barr, J.F. Donoghue and D. Seckel,Phys. Rev. D57, 5480 (1998).

N. Arkani-Hamed and S. Dimopoulos, JHEP 0506, 073(2005); G.F. Giudice and A. Romanino, Nucl. Phys. B699, 65(2004) [erratum: B706, 65 (2005)]. July 27, 2006 11:28

en.wikipedia.org/wiki/Supersymmetry - 52k - Cached - Similar pages

en.wikipedia.org/wiki/Grand_unification_theory - 39k - Cached - Similar pages

In cosmology, the Planck epoch (or Planck era), named after Max Planck, is the earliest period of time in the history of the universe, en.wikipedia.org/wiki/**Planck_epoch** - 23k - Cached - Similar pages

L. Shapiro and J. Sol`a, Phys. Lett. B 530, 10 (2002);

E. V.Gorbar and I. L. Shapiro, JHEP 02, 021 (2003); A. M. Pelinson,

L. Shapiro, and F. I. Takakura, Nucl. Phys. B 648, 417 (2003).

A. Starobinsky, Phys. Lett. B 91, 99 (1980).

G. F. R. Ellis, J. Murugan, and C. G. Tsagas, Class. Quant. Grav.21, 233 (2004).

H. V. Peiris et al., Astrophys. J. Suppl. 148, 213 (2003).

D. N. Spergel et al., astro-ph/0603449.

Vilenkin, Phys. Rev. D 32, 2511 (1985).

A. Starobinsky, Pis'ma Astron. Zh 9, 579 (1983).

A.H. Guth, Phys. Rev. D23, 347 (1981).

A.D. Linde, Phys. Lett. B108, 389 (1982); A. Albrecht, P.J.

Steinhardt, Phys.Rev. Lett. 48, 1220 (1982).

A.D. Linde, Phys Lett. B129, 177 (1983).

N. Makino, M. Sasaki, Prog. Theor. Phys. 86, 103 (1991);

D. Kaiser, Phys. Rev.D52, 4295 (1995).

H. Goldberg, Phys. Rev. Lett. 50, 1419 (1983).

E. Kolb and M. Turner, The Early Universe (Addison-Wesley, Reading, MA,1990).

W. Garretson and E. Carlson, Phys. Lett. B 315, 232(1993); H. Goldberg, hep-ph/0003197.

Eddington, A. S., The Internal Constitution of the Stars (Cambridge University Press, England,1926), p. 16

Duncan R .C. & Thompson C., Ap.J.392, L 9 (1992).

Thompson , C, Duncan , R .C ., Woods , P., Kouveliotou , C ., Finger , M.H. & van Parad ij s , J .,ApJ, submitted , astro-ph /9908086, (2000).

Schwinger , J .,Phys. Rev.73, 416L (1948)

Carlip, S.: Quantum gravity: a progress report. Rept. Prog. Phys. 64, 885 (2001).arXiv:gr-qc/0108040

Kerr,R.P.: Gravitational field of a spinning mass as an example of algebraically special metrics.

Phys. Rev. Lett. 11, 237–238 (1963)

Bekenstein, J.: Black holes and the second law. Lett. Nuovo Cim. 4, 737–740 (1972)

Bardeen, J.M., Carter, B., Hawking, S.: The four laws of black hole mechanics. Commun.

Math. Phys. 31, 161–170 (1973)

Tolman, R.: Relativity, Thermodynamics, and Cosmology. Dover Books on Physics Series.

Dover Publications, New York (1987)

Oppenheimer, J., Volkoff, G.: On massive neutron cores. Phys. Rev. 55, 374–381 (1939)

Tolman, R.C.: Static solutions of einstein's field equations for spheres of fluid, Phys. Rev. 55, 364–373 (1939)

Zel'dovich Y.B.: Zh. Eksp. Teoret. Fiz.41, 1609 (1961)

Bondi, H.: Massive spheres in general relativity. Proc. Roy. Soc. Lond. A281, 303–317 (1964)

Sorkin, R.D., Wald, R.M., Zhang, Z.J.: Entropy of selfgravitating radiation. Gen. Rel. Grav. 1127–1146 (1981)

Newman, E.T., Couch, R., Chinnapared, K., Exton, A., Prakash, A., et al.: Metric of a rotating,charged mass. J. Math. Phys. 6, 918–919 (1965)

Ginzburg, V., Ozernoi, L.: Sov. Phys. JETP 20, 689 (1965)

Doroshkevich, A., Zel'dovich, Y., Novikov I.: Gravitational collapse of nonsymmetric and rotating masses, JETP 49 (1965)

Israel, W.: Event horizons in static vacuum space-times. Phys. Rev. 164, 1776–1779 (1967)

Israel, W.: Event horizons in static electrovac space-times. Commun. Math. Phys. 8, 245–260 (1968)

Loop quantum gravity does provide such a prediction [363, 364], and it disagrees with the semiclassical

Carter, B.: Axisymmetric black hole has only two degrees of freedom. Phys. Rev. Lett. 26, 331–333 (1971)

Penrose, R.: Gravitational collapse: the role of general relativity. Riv. Nuovo Cim. 1, 252–276 (1969)

Christodoulou, D.: Reversible and irreversible transformations in black hole physics. Phys. Rev. Lett. 25, 1596–1597 (1970)

Christodoulou, D., Ruffini, R.: Reversible transformations of a charged black hole. Phys. Rev. D4, 3552–3555 (1971)

Hawking, S.: Particle creation by black holes. Commun. Math. Phys. 43, 199–220 (1975)

Klein, O.: Die reflexion von elektronen an einem potential sprung nach der relativistischen dynamik von dirac. Z. Phys. 53, 157 (1929)

Wald, R.M.: General Relativity. The University of Chicago Press, Chicago (1984)

Hawking, S.W.: Black hole explosions. Nature 248, 30–31 (1974)

Hawking, S., Ellis, G.: The large scale structure of space-time. Cambridge University Press, Cambridge (1973)

Carter, B.: Black hole equilibrium states, In Black Holes—Les astres occlus. Gordon and Breach Science Publishers, (1973)

Hawking, S.W.: Gravitational radiation from colliding black holes. Phys. Rev. Lett. 26, 1344–1346 (1971)

Hawking, S.: Black holes in general relativity. Commun. Math. Phys. 25, 152–166 (1972)

Bekenstein, J.: Extraction of energy and charge from a black hole. Phys. Rev. D7, 949–953 (1973)

Bekenstein, J.D.: Black holes and entropy. Phys. Rev. D7, 2333–2346 (1973)

Hawking, S.: Quantum gravity and path integrals. Phys. Rev. D18, 1747–1753 (1978)

Gross, D.J., Perry, M.J., Yaffe, L.G.: Instability of flat space at finite temperature. Phys. Rev. D25, 330–355 (1982)

Unruh, W.G., Wald, R.M.: What happens when an accelerating observer detects a rindler particle. Phys. Rev. D29, 1047–1056 (1984)

Bekenstein, J.D.: Auniversal upper bound on the entropy to energy ratio for bounded systems. Phys. Rev. D23, 287 (1981)

Unruh,W.,Wald, R.M.: Acceleration radiation and generalized second law of thermodynamics. Phys. Rev. D25, 942–958 (1982)

Unruh, W., Wald, R.M.: Entropy bounds, acceleration radiation, and the generalized second law. Phys. Rev. D27, 2271–2276 (1983)

Image : MPI for gravitational physics / W.Benger-zib

Tomilin,K.A., (1999). "Natural Systems Of Units: To The Centenary Aniniversary Of The Planck Systems", 287-296

Sivaram, C. (2007). "What Is Special About the Planck Mass"? arXiv:0707.0058v1

H. Georgi and S.L. Glahow. (1974) "Unity Of All Elementary-Particle Forces". Phys. Rev. Letters 32, 438

Luigi Maxmilian Caligiuri, Amrit Sorli. Gravity Originates from Variable Energy Density of Quantum Vacuum. American Journal of Modern Physics. Vol. 3, No. 3, 2014, pp. 118-128. doi: 10.11648/j.ajmp.20140303.11

Philip J. Tattersall,(2018) Quantum Vacuum Energy and the Emergence of Gravity. doi:10.5539/apr.v10n2p1

H. E. Puthoff (1989) Gravity as a zero-point-fluctuation force PHYSICAL REVIEW A VOLUME 39, NUMBER 5

Balungi Francis (2018) "Quantum Gravity in a Nutshell1" Book.

E.Verlinde (2016) Emergent Gravity and the Dark Universe, arXiv:1611.02269v2[hep-th]

S.Hossenfelder (2018) The Redshift-Dependence of Radial Acceleration: Modified gravity versus particle dark matter, arXiv:1803.08683v1[gr-qc]

Robert J. Scherrer (2004) Purely kinetic k-essence as unified dark matter, arXiv:astro-ph/0402316v3

J.S.Farnes (2018), Aunifying theory of dark energy and dark matter: Negative masses and matter creation within a modified ΛCDM framework, arXiv:1712.07962v2[physics.gen-ph]

Gustav M Obermair (2013), Primordial Planck mass black holes (PPMBHs) as candidates for dark matter? J. Phys:conf.Ser.442012066

V.Cooray etal…(2017), An alternative approach to estimate the vacuum energy density of free space, doi:10.20944/preprints201707.0048.v1

M.Milgrom, (1983) A modification of the Newtonian dynamics: Implications for galaxies, Astrophys.J.270, 371.

Acknowledgments

This book would never have been completed without the patience and dedication of my wife, Wanyana Ritah. She performed the wonderful and difficult task of editing major parts of the book and helped in researching many details necessary to complete it.

I wish to thank several colleagues for their help and extensive comments on the manuscript, including Lee Smolin, Carlo Rovelli, Sabine Hossenfelder, Jim Baggot and Viktor Toth. I also thank my colleagues Harvey Brown, Paul Frampton, Stacy McGaugh, and Lee Smolin for helpful comments. I particularly thank a total of 200 online physics friends and SUSP science foundation members, for a careful reading of the manuscript. Many graduate students have contributed over the years to developing my Quantum theory of gravity.

I also wish to thank my editors, at SUSP science Foundation for their enthusiasm and support. Finally, I thank our family for their patience, love, and support during the four years of working on this book.

About the Author

Balungi Francis was born in Kampala, Uganda, to a single poor mother, grew up in Kawempe, and later joined Makerere Universty in 2006, graduating with a Bachelor Science degree in Land Surveying in 2010. For four years he taught in Kampala City high schools, majoring in the fields of Gravitation and Quantum Physics. His first book, "Mathematical Foundation of the Quantum theory of Gravity," won the Young Kampala Innovative Prize and was mentioned in the African Next Einstein Book Prize (ANE).

He has spent over 15years researching and discovering connections in physics, mathematics, geometry, cosmology, quantum mechanics, gravity, in addition to astrophysics, unified physics and geographical information systems . These studies led to his groundbreaking theories, published papers, books and patented inventions in the science of Quantum Gravity, which have received worldwide recognition.

From these discoveries, Balungi founded the SUSP (Solutions to the Unsolved Scientific Problems) Project Foundation in 2004 - now known as the SUSP Science Foundation. As its current Director of Research, Balungi leads physicists, mathematicians and engineers in exploring Quantum Gravity principles and their implications in our world today and for future generations.b

Balungi launched the Visionary School of Quantum Gravity in 2016 in order to bring the learning and community further together. It's the first and only Quantum Gravity physics program of its kind, educating thousands of students from over 80 countries.

The book "Quantum Gravity in a Nutshell1", a most recommend book in quantum gravity research, was produced based on Balungi's discoveries and their potential for generations to come. Balungi is currently guiding the Foundation, speaking to audiences worldwide, and continuing his groundbreaking research.

Contact Balungi Francis at the following address:

Email: balungif@gmail.com

bfrancis@cedat.mak.ac.ug

BILL STONE SERVICES

Tel: +256703683756

+256777105605

www.ingramcontent.com/pod-product-compliance
Lightning Source LLC
Chambersburg PA
CBHW031603210526
45464CB00004B/1412